D0472190

World Changing Ideas

# World Changing Ideas

Richard Myers & Bob Isherwood

Saatchi & Saatchi
375 Hudson Street
New York
NY 10014-3620
USA

Published by
Palazzo Editions Ltd
15 Gay Street
Bath BA1 2PH



Publishing manager: Colin Webb
Project editor: Sonya Newland
Coordinator for Saatchi & Saatchi: Norma Clarke
Design Director Saatchi & Saatchi: Roger Kennedy



A CIP catalogue record for this book is available from the British Library

ISBN 0 9553046 0 1

Printed and bound in Italy by Conti Tipocolor. Printed on 130gsm Arctic the Volume, a Forest Stewardship Council (FSC)-certified paper containing wood from well-managed forests. For more information, see www.fsc.org

## Contents

Foreword 7

Introduction 10

1 The Natural World 16

2 The Language World 52

3 The Knowledge World 70

4 The Technology World 84

5 The Medical World 138

6 The Disability World 166

7 The Developing World 208

Awards and Judges 236

Index 255

# Foreword



As an astronaut, I've enjoyed a highly privileged view of Earth as it hurtles through space on its orbit around the Sun.

For the moment, it's the only place in the Universe we can call home, although I'm convinced this won't always be the case.

Our world isn't a single entity, of course, but a composite of many smaller worlds. From continents to nation states, right through to the world of every individual person.

Even the most blithely optimistic person would not claim that everything about all of these worlds is ideal.

Although it's tempting to 'measure' issues and describe them as 'major' or 'minor' or 'global' or 'local', they should really all be seen as equal. I say this because every one of them has an impact, whether it's on the well-being of an individual or of the whole planet.

Issues need to be resolved, and issues are resolved by ideas.

The fortunate thing is, the vast majority of the human race is wonderfully resourceful, inventive and ingenious. How else would we have been able not simply to survive for over two million years, but to develop to such a level of technical and cultural sophistication?

As a species we can't resist exploring and we can't resist solving problems and chasing improvements.

The changes that great ideas instigate can travel in two directions. Either from the individual's world through to the whole world, or the other way round.

Think of the pebble in the pond where the ripples radiate outwards across the water, and then imagine the ripples starting at the outer edge and spreading inwards to the pebble. Equally, great ideas which ostensibly tackle one issue, can actually produce, organically, a multiple of benefits across a range of problems.

When you look at the incredible ideas in this book, you'll see exactly what I mean.

When Saatchi & Saatchi invited me to judge their inaugural Award in 1998, the proposition appealed to me as an inventor, as an explorer, and as a human being.

It's vital that ideas with the potential to be world changing are identified and championed.

The ideas collected here tell me I was right to follow my instincts about the Award and accept the invitation.

Ultimately, I believe there isn't a single problem we face now or will face in the future that we can't solve with ideas.

*Buzz Aldrin*

# Introduction



Progress relies entirely on one fuel. Ideas. Without them, there's stagnation at best. Regression at worst.

And so many big ideas revolve around communication. Why? For the simple reason that communication lies at the heart of every human endeavour and every human life. It isn't hysterical to suggest that our planet's very survival relies, ultimately, on our ability and willingness to communicate.

Innovations in communication are therefore as important and as impactful as any invention you care to think of. Yes, the wheel counts.

Reflect on the significance of how we communicate with each other as individuals.

Expand that to communities, and then to nations and then to planets.

Consider how we communicate with our immediate environment. And how it's impossible for some people to communicate with theirs because of a disability.

And how it's impossible for other people to communicate because of their lack of education or the technology most of us take for granted.

The very fact that you're reading this book is a demonstration of the differences between the developing world and the developed world.

You can read, for a start.

If it's dark, it doesn't matter. You have electric light to read by. If your eyesight is poor, that doesn't matter either. You can wear glasses or contact lenses to correct whatever sight problems you have.

You also have a broad base of knowledge, which means you can understand the content of this book.

On 14 August 2002, the population of the United States' eastern seaboard was given an unwelcome reminder of how much we take for granted, when their electricity supply failed. For two billion people in the developing world this is everyday life.

But it isn't only the developing world that needs world changing ideas.

The whole world does.

Because the adjective that can always be applied to the world is 'imperfect'. We'll never run out of things that need changing for the better.

Even the cleverest, latest technology has a habit of creating new problems.

For example, how do you recharge your mobile- or cell-phone battery when there's no mains electricity?

How do you make your fingers small enough to use ever-more miniaturized keyboards?

If either of these is a problem for you, then having a solution is world changing for you.

In many cases, changing the world of the individual will see the benefit expand and multiply to change the whole world, or at least a significant proportion of it.

The benefit cascade of a world changing idea can be quite astonishing.

You would think that the benefit of an affordable form of electric light for the developing world would be just that. Light.

Not so.

The health, environmental, economic and security benefits that roll out from the lighting are immeasurable.

Fortunately, there are people who view this imperfect world as a world of opportunity.

These are the innovators, the inventors who, in the words of one of them, American David Levy, 'go looking for trouble'.

They are people of crackling ingenuity.

And they don't all wear sandals, thick glasses and white coats, or work away in wooden shacks or garden sheds that they frequently destroy in unscheduled explosions.

This overlooks the army of brilliant minds populating the world's top academic institutions.

Most have exceptional determination and focus.

Most have extraordinarily creative minds.

Most are the unusual blend of dreamers and doers.

Most care passionately about the world around them.

And all of them can't help themselves. They have to invent.

Necessity isn't the only mother of invention. Vision is, too.

Inventors have to be thick-skinned. Their single-mindedness can generate ridicule. Their vision can generate scorn. By definition, if an idea is ahead of its time, its benefit will be invisible to anyone who can't see beyond today.

Rejection is also an occupational hazard. And so is daylight robbery. The anarchist's anti-capitalist view that 'property is theft', has been modified by some fine capitalist corporations to 'intellectual property is there to be thieved'.

Although an inventor's primary motivation may not be making money, they do need funding. First to protect their idea with patents and then to make it happen.

The majority can't afford to sue corporations who steal their ideas. The majority of corporations know this.

It could be argued that as long as the world benefits from the invention, does it matter, ultimately, who brings it to us?

Even the most philanthropic inventor would probably find that hard to swallow.

There is, however, the scary thought that there may be thousands of incredible world changing ideas out there that we simply don't know about. And worse, that we never will.

This book is about the world changing ideas that have been spotlighted by being entered into a global competition.

The competition is run by Saatchi & Saatchi, a company founded in 1970 as an advertising agency, which has now evolved into an ideas company.

The competition was launched in 1998, as the Saatchi & Saatchi Award for Innovation in Communication. You'll read how, right from the start, the competition attracted brilliant entries. And some truly inspiring judges.

It also became clear right from the start just how significant communication is to the well-being of the individual and of the whole planet. The diversity of the entries bears out the earlier assertion about communication lying at the heart of every human endeavour and every human life.

The competition is now called the Saatchi & Saatchi Award for World Changing Ideas (for Innovation in Communication).

But the Award is not the hero.

The ideas are.

*Richard Myers & Bob Isherwood*

1

# The Natural World

# Don't you want to know when a twister's coming?

Getting out of the way of a column of warm air rotating at up to 480 km/h (300 mph) and heading towards you is a really good idea. In the US alone, tornadoes cause millions of dollars' worth of damage to property and infrastructure every year. The cost in human life is high too, with between 50 and 200 people killed. But getting out of the way of a tornado or finding shelter that's robust enough to withstand this most violent of atmospheric storms, relies on an effective warning system.

Doppler radar has helped in recent years by detecting air motion associated with an approaching twister.

Unfortunately, Doppler radar can't determine when a tornado is in its most lethal position – when it's in touch with the ground.

But it's been discovered that when a tornado makes contact with the ground it produces considerable seismic signals. This is the basis for Dr Frank Tatom's invention, Seismic Detection of Tornadoes.

He envisages a network of Seismic Tornado Detectors that can pick up signals as much as 40 km (25 miles) away. They would be used alongside the Doppler radar to create a potentially life-saving alarm system.

Dr Tatom's network of detectors would be made up of a series of geophone components, sensors that pick up any vibrations on the ground. These are buried below the surface 16 km (10 miles) apart and linked to a data control centre where the seismic signals would be processed. Quiet locations, like churchyards, are ideal for the detectors. Anywhere near railway lines should be avoided as the rumbling caused by freight trains could trigger the detectors.

Dr Tatom's system is designed for government agency or private commercial use. When a tornado was detected on the ground, the data control centre would broadcast a special warning signal that would trigger alarms in occupied structures, including office buildings, apartment houses, trailer parks and individual homes, where the alarm and microchip would function in a similar way to a smoke detector.

*Top left* Shallow hole being dug for geophone component of detection instrument package, nicknamed 'Snail'.

*Top right* Snail handler presses geophone into hole.

*Above* Handler removes Snail case from chase vehicle.

*Above right* Handler deploys Snail in open field as tornado approaches from left.

*Right* Handler leaves field after deploying Snail.

Families with the system would get between two and four crucial minutes to move to the safest part of their house.

Dr Tatom has set up a commercial company, VorTek, to develop his invention further. He's chosen to base it in Huntsville, in the tornado prone state of Alabama. He can see his Seismic Detection System being particularly useful there, in the south-eastern United States, where the low cloud base, uneven terrain and lots of trees tend to prevent timely visual sighting.

And when you live in a tornado-prone region, there are certain things it's really helpful to know.

**Tornadoes and hurricanes** Thunderstorms give birth to tornadoes. A funnel-shaped column of violently whirling air forms under a thundercloud and heads for the ground. This column is known as the mesocyclone and it can destroy everything it meets, tearing houses apart and tossing vehicles into the air. Tornadoes are usually about 300 m (1,000 ft) across and travel over the ground at about 50 km/h (31 mph), with winds between 160–480 km/h (100–300 mph).

By contrast, hurricanes – the name given to tropical cyclones north of the Equator – are vast. They form when the temperature of the ocean is at least 26.5°C (80°F) down to at least 50 m (150 ft). Enormous spirals of dense clouds develop, up to 500 km (300 miles) across, with winds inwardly spiralling around at up to 320 km/h (200 mph).

In the northern hemisphere, the wind spirals counter-clockwise, and in the southern hemisphere, clockwise.

An oddity in all this violent activity is the eye of the hurricane, an area about 30 km (19 miles) across at the storm's centre, where it is calm, with little wind or cloud.

# Putting extinction on ice.

Frozen Ark Project
FINALIST, 2005

The rate of extinction of animal species is reckoned by some scientists to be on a scale equal to the last five great extinctions. In fact, they are referring to this plunge in biodiversity as Earth's 'sixth mass extinction'.

Other scientists say only the three greatest environmental disasters in the entire history of the Earth killed species at a greater rate.

The United Nations Environment Programme predicts that by 2075 few locally endemic species will remain. One quarter of the world's mammals and 12 per cent of its bird species face extinction within 30 years, and the invertebrates

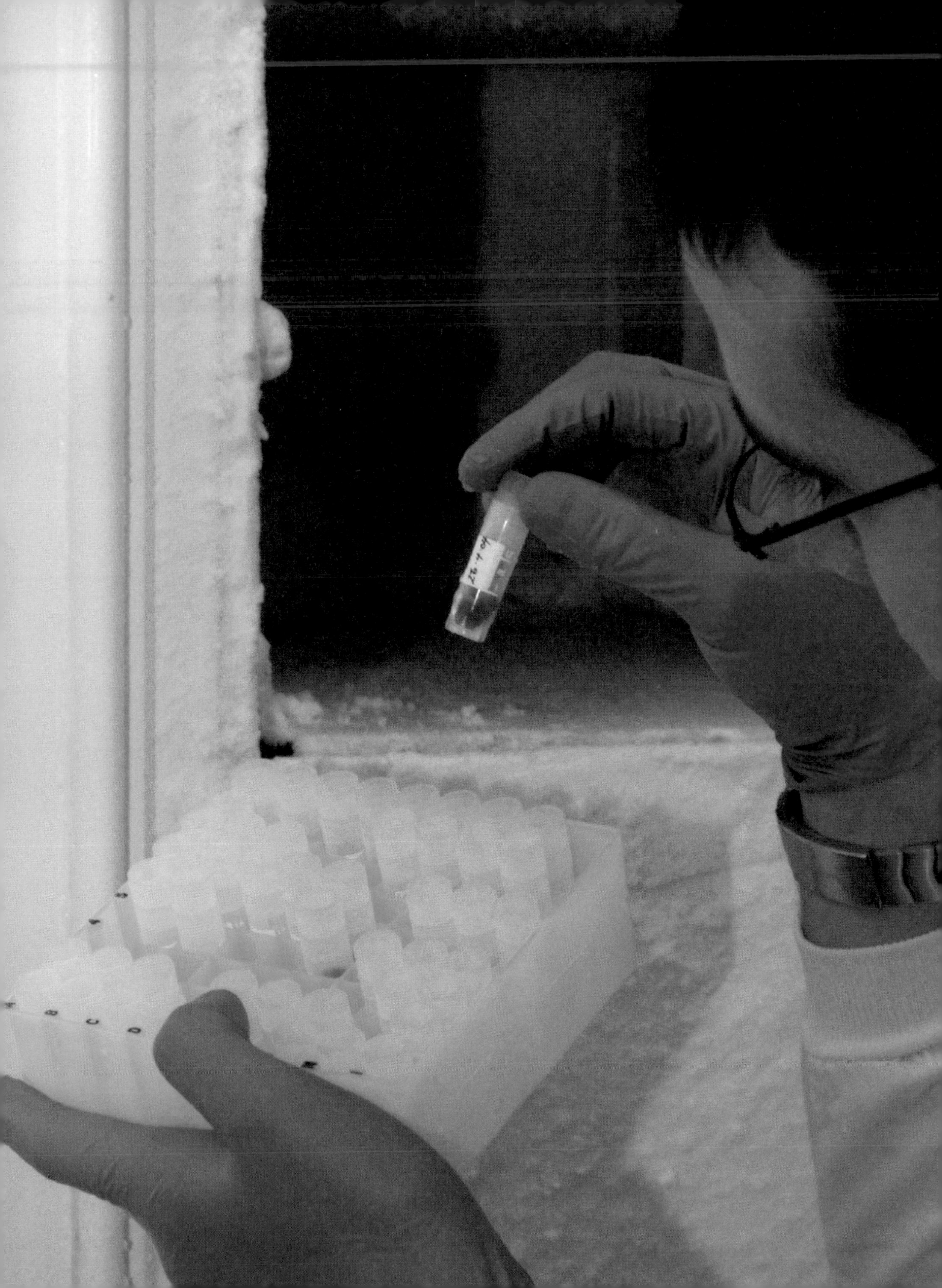

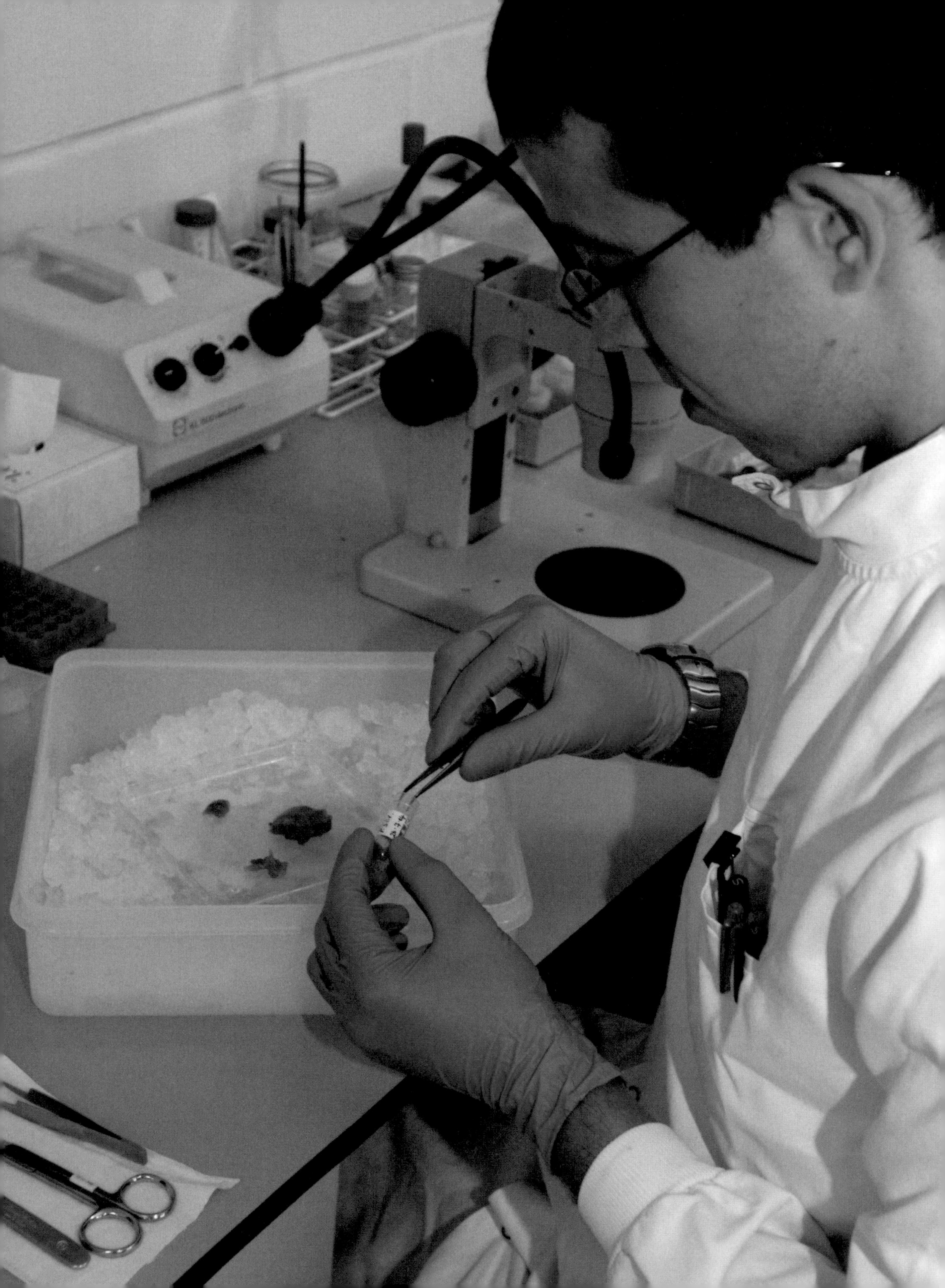

will cope no better. All this, despite our best efforts at conservation.

Ten thousand animal species are currently endangered and this threat to Earth's biodiversity is, ultimately, a threat to our own survival.

In history, ice may have played the villain in killing off species. Now it could be the hero.

The idea of the Frozen Ark Project is to save the DNA of each endangered species before it becomes extinct, by storing samples at minus 196°C (minus 321°F). The initial cost is around £300 ($535) per species, excluding the cost of collection, and the sample could stay intact for tens of thousands of years, perhaps more.

The DNA of all animal species on Earth could be stored in an average-sized house.

A DNA sample contains the organism's blueprint, and can give scientists a vast amount of information about its relationships, evolution, genetics, development, diseases and ecology. Importantly, it can be used to introduce valuable genetic variation into species still being bred in captivity.

The advances in DNA technology suggest that it may even be possible in the future to reconstruct the whole animal, particularly if viable cells were available. So the Frozen Ark intends to freeze such cells as well, although the process is more costly.

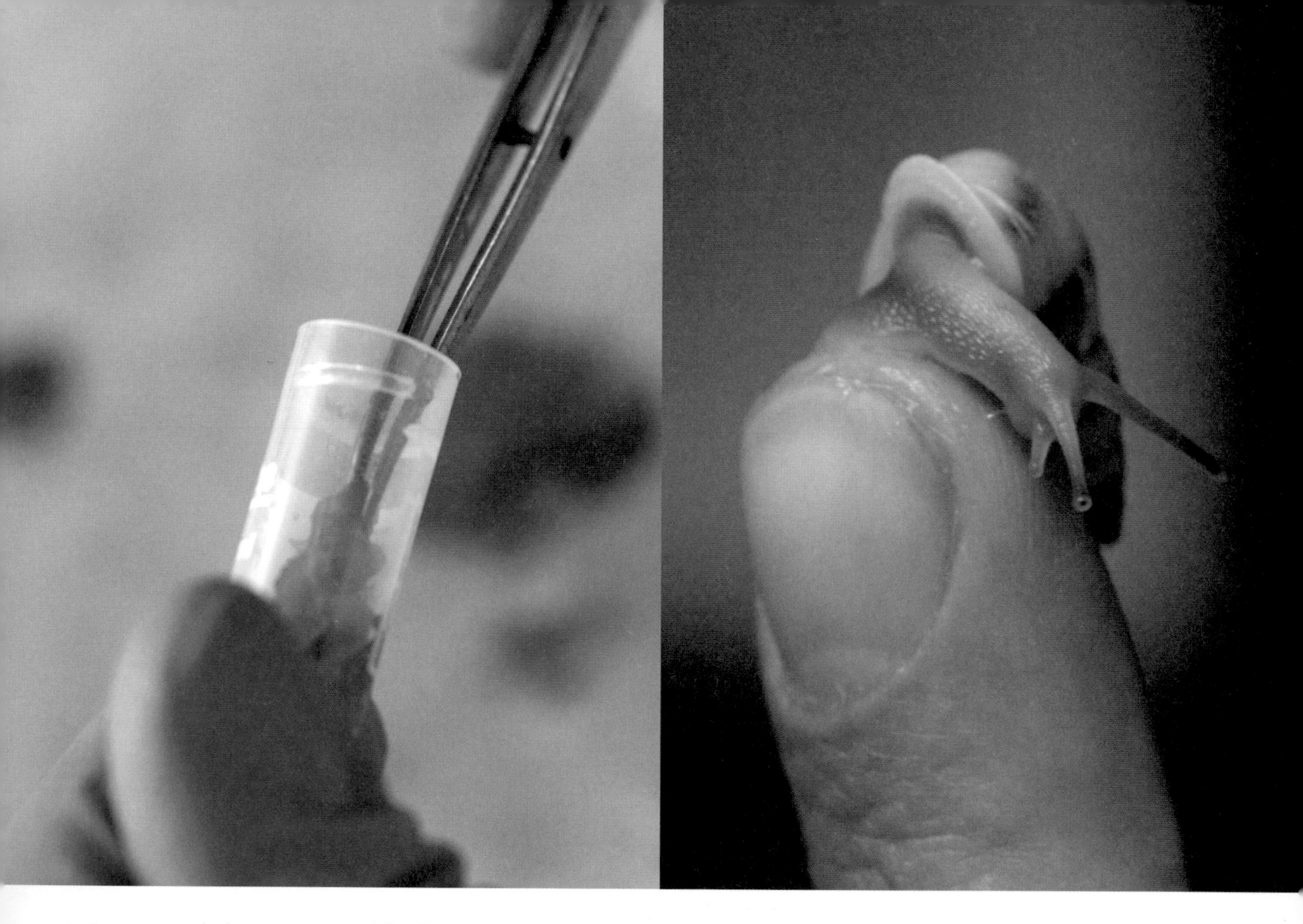

A tissue sample being prepared for freezing.

Polynesian tree snail (*Partula suturalis*). Extinct in the wild.

There's no time to lose for the Frozen Ark Project and priority is being given to animals at risk of dying out in the next five years. Rescuing the information may save the species.

The Project is being initiated in the United Kingdom, at the University of Nottingham, with a growing number of worldwide consortium members, including the Natural History Museum and The Institute of Zoology at London Zoo, The American Museum of Natural History, San Diego Zoo, The Animal Gene Storage Research Centre of Australia, and The Endangered Wildlife Trust of South Africa.

Centres will collect and store their own native specimens and act as repositories for duplicate specimens as an insurance against damage or loss. In fact, a key component of the Project will be the exchange of information and expertise between the Frozen Ark centres.

Like its biblical equivalent, the Frozen Ark will hold DNA from both male and female animals. But unlike its biblical equivalent, this Ark has three 'Noahs' – Professor Bryan C. Clarke, Dame Anne McLaren and Dr Ann Clarke, who recognized that conservation alone is not enough and that a back-up is vital.

DNA DNA, deoxyribonucleic acid, is the molecule that carries all the blueprints of life.

As a human being, each cell in your body has the same set of about 30,000 genes. (An adult human has about 10 million million cells.) DNA works by telling a cell how to make the proteins that keep it alive and growing. All these genes add up to your genome, which is a complete set of instructions for building you from scratch. Your genes are arranged along long, thin, paired structures called chromosomes. Humans have 23 pairs (46 chromosomes) in every cell. Sperm and egg cells are different. They each have 23 unpaired chromosomes. When a sperm fertilizes an egg cell, each parent contributes 23 unpaired chromosomes. The baby, therefore, has its own unique set of 46 chromosomes. In fact, the chances of two people, other than identical twins, having identical DNA has been estimated at one in 738,000,000,000,000.

At the same time, almost a third of human genes are similar to those of a lettuce.

# It's a disaster.
# A hospital built in 24 hours.

When your world has been changed by a natural or man-made disaster, you need it to change again for the better.

For a variety of reasons, it's estimated that there are over 35 million refugees worldwide, and at the moment, available shelter is either soft-skinned (e.g. tents) or expensive, difficult-to-transport prefabricated structures.

The inadequate protection provided by tents is magnified in Afghanistan, for instance, where wind damage can leave some tents unusable in less than three weeks.

The United Nations coordinator for Afghanistan has made the observation that, 'while starvation occurs over a period of weeks, death from exposure can occur in a single night'.

Now the good news.

Concrete Canvas is the invention of two postgraduate engineering design students from London's Royal College of Art – Peter Brewin and William Crawford.

They say that their invention 'provides the infrastructure necessary for aid agencies to communicate and operate effectively anywhere in less than 24 hours'. And that with shelter and medical facilities it is possible to rebuild shattered communities from day one of a crisis.

Described as a 'building in a bag', a Concrete Canvas Shelter only needs water and air for its construction. One untrained person can put the structure together in under 40 minutes and it will be ready to use in 24 hours. More rapid variants are under development, reducing this time to under 12 hours.

The structure has two elements: a cement-impregnated fabric (concrete cloth) that is bonded to the outer surface of an inflatable plastic inner.

The use of inflation is key in creating a surface optimized for compressive loading. This allows thin-walled concrete structures to be formed that are both robust and lightweight.

The unit is delivered folded in a sealed plastic sack, weighing 500 kg (1,100 lb). It's small enough to fit on a pick-up truck or to be air dropped.

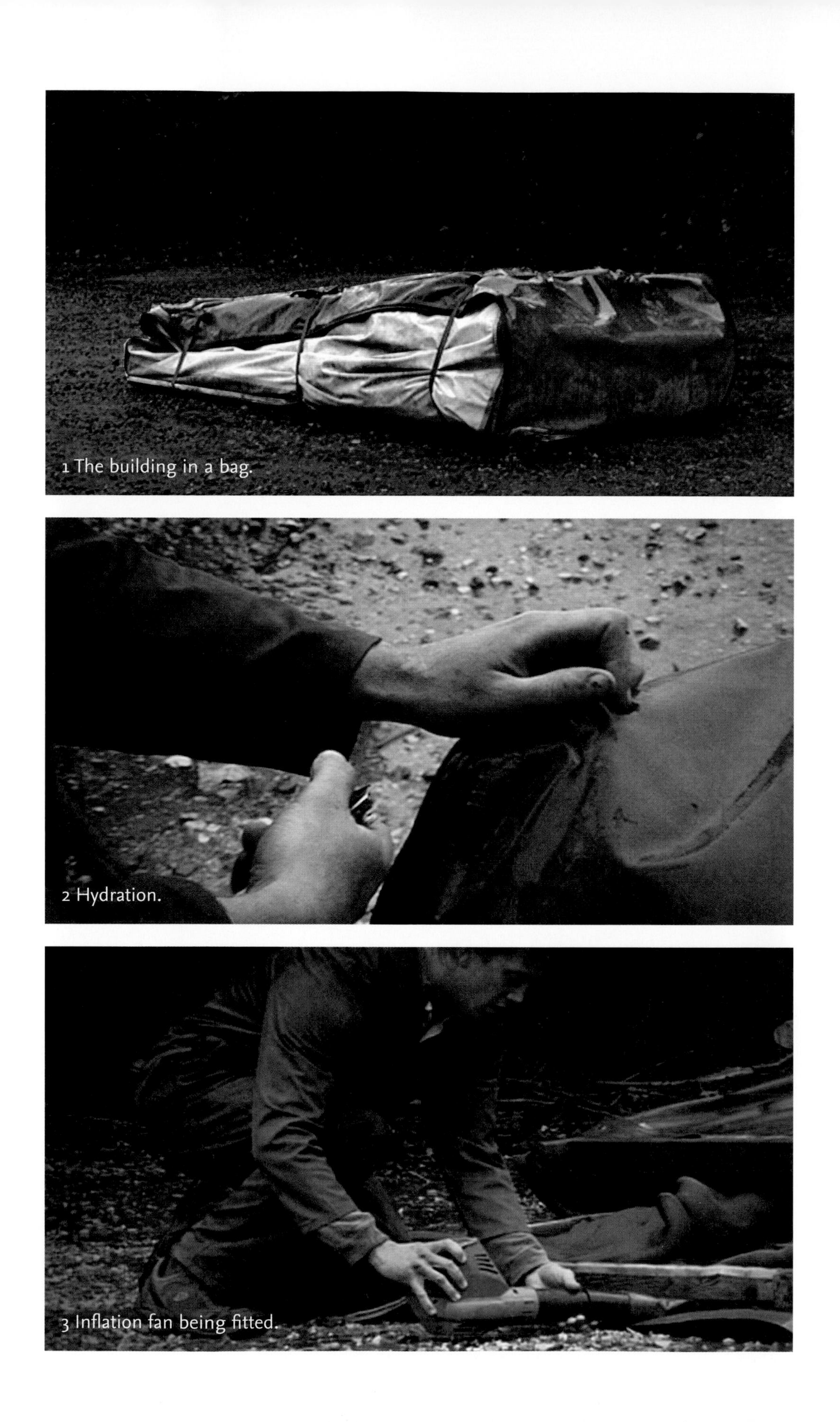

1 The building in a bag.

2 Hydration.

3 Inflation fan being fitted.

4 Shelter half inflated.

5 Shelter fully inflated.

6 The finished Shelter.

Step one is hydration. The sack is positioned and filled with 145 litres (32 gallons) of any water, apart from sewage or seawater. (This is equivalent to the UN's recommended daily water ration for 12 people.) The water doesn't have to be measured out because the volume of the sack controls the ratio of cement to water. After 15 minutes the cement is hydrated and the sack is cut open, which then forms part of the ground sheet.

Step two is inflation. The Shelter is delivered with its own inflation pack, a battery-driven fan. The hydrated cement structure is unfolded and the fan pumps air into the plastic inner liner to inflate the structure.

Step three is setting. The concrete cloth sets in the shape of the inflated plastic liner, and less than 24 hours later the 16 m² (172 ft²) Shelter is ready to use. Parts of the plastic liner are left clear of concrete cloth so that doors and ventilation holes can be easily cut out once the cement sets. Because of the way the doorways are designed, it's possible for several Shelters to be linked together.

Concrete Canvas not only provides shelter for victims and refugees. It can be the vital initial infrastructure for aid agencies and for troops providing care and protection. It can also provide a safer store for supplies. In another context, Concrete Canvas could be used for field operating hospitals, closer to the battle front than conventional structures could be sited.

Concrete Canvas is a virtuous contradiction: it can be rapidly deployed and erected, yet it's durable. It has the added virtues, in a crisis, of being medically sterilizable, secure, insulating and strong, and it has the structural qualities to perform well during earthquakes.

A Concrete Canvas Shelter is near-instant infrastructure that could last over 10 years if it had to. In some cases, that could prove essential.



**Famous refugees** A number of highly successful individuals across a wide range of activities have in common the fact that they were refugees.
Art: Piet Mondrian, Marc Chagall, Jacob Epstein, Lucien Freud and Peter Carl Fabergé.
Business: Michael Marks, the co-founder of the UK retailers Marks & Spencer.
Design: Sir Alec Issigonis, the designer of the original Mini.
Music: Sir Georg Solti and Arnold Schoenberg.
Politics: Madeleine Albright, Henry Kissinger, Leon Trotsky and Karl Marx.
Psychology: Sigmund Freud.
Religion: Rabbi Immanuel Jakobovits, the Dalai Lama.
Science: Albert Einstein, Nobel Prize-winning physicist Max Born.
TV & Film: Billy Wilder.
Writing & Publishing: Isabel Allende, Joseph Conrad, Anne Frank, Paul Hamlyn, Thomas Mann, Vladimir Nabokov.

# Predicting the unpredictable.

Global Earthquake Monitoring System (GEMS)

FINALIST, 2002

At the time of writing this, television screens and newspapers were full of images of devastation, caused by the South Asia earthquake centred around Kashmir and Pakistan. Less than 10 months earlier, the 2004 Sumatra–Andaman earthquake caused the Indian Ocean tsunami, killing 300,000 people and leaving millions homeless and destitute.

Before that, the 2003 Bam earthquake in Iran killed 34,000 people. Forty thousand died in Bhuj, India in 2001, and there were 17,000 deaths caused by the 1999 Izmit earthquake in Turkey. The 1995 Kobe earthquake in Japan claimed 6,000 lives and caused £240 billion ($450 billion) of damage.

The total cost of 10 years' earthquakes is more than £1,000 billion ($1,900 billion). The fact of the matter is, earthquakes are horribly unpredictable and vastly expensive.

The renowned physicist Richard Feynman once said: 'We understand the distribution of matter in the interior of the Sun far better than we understand the interior of the Earth'.

Professor Stuart Crampin, based at Edinburgh University's School of GeoSciences, puts it this way: 'Despite over 200 years of geology and 100 years of geophysics, we know remarkably little about how rocks deform a few metres beneath our feet.'

However, Professor Crampin has developed a radical new approach to forecasting earthquakes. 'Previously, people studied the earthquake source. Our breakthrough was that we found we could monitor the build-up of stress before large earthquakes at substantial distances from the eventual earthquake. With GEMS, the Global Earthquake Monitoring System, you could forecast all damaging earthquakes worldwide, and there are possibilities of reducing damage by releasing stress in non-vulnerable areas by injecting fluids at high pressures into the stressed rock.'

Professor Crampin's research has led academics to understand the importance of the tiny cracks in underground rocks, which contain fluid and are aligned by the stress field.

11 January 2005. The aftermath of the Indian Ocean tsunami at Banda Aceh, Indonesia.

These microcracks in the Earth's crust are highly compliant to small changes. Any minuscule variations in the initial stress may lead to substantially different behaviour (deterministic chaos) when the rocks eventually fail in earthquakes. (The chaos is why studying the source is probably not useful for prediction.)

'There is a build-up of stress before large earthquakes. We can monitor these changes and recognize when the microcracks approach failure in imminent earthquakes. If a large earthquake was expected offshore, it could cause a tsunami,' Professor Crampin explains.

He and his team have tested their theories using small 'swarms' of earthquakes as the signal source and have seen the effects, with hindsight, before some 15 earthquakes worldwide. On one occasion, they predicted a damaging earthquake in south-west Iceland before it happened.

To forecast all damaging earthquakes worldwide in GEMS calls for a network of borehole Stress-Monitoring Sites around the Earth at intervals of 400 km (250 miles), so some 1,500 SMSs would be needed to create the complete grid.

Because his findings undermine received wisdom, Professor Crampin has struggled to get them embraced by established Earth scientists.

'These developments are new, controversial, not fully understood, difficult to get accepted and, consequently, difficult to get funded,' he admits. Also, the estimated £5.3 billion ($10 billion) cost of a Global Earthquake Monitoring System looks high. Until you relate it to the cost in human life and property caused by one of the world's deadliest natural phenomena.

Professor Crampin's email is scrampin@ed.ac.uk.

**The father of geology** By coincidence, James Hutton, the 'father of geology' also came from Edinburgh.

He lived between 1726 and 1797, during a time when most people based their belief that the Earth was only 6,000 years old on the Bible.

In *A Theory of the Earth*, in 1788, Hutton proposed, controversially, that the Earth was 'infinitely' old.

His theory of Uniformitarianism suggested that events in geological history could be explained in terms of processes that still apply today.

For example, the kind of current in a river that produces a particular settling pattern in a sand bed today, must have been operating millions of years ago, if that same pattern is visible in ancient sandstones.

# A corny idea to tackle pollution.

Plantic®
FINALIST, 2005

How many ideas can you think of that make a problem literally disappear?

Here's one.

Plantic® – Plastic From Plants – is a cutting-edge substitute for plastic packaging made from cornstarch. It provides the protection, presentation and economies that traditional plastic is expected to deliver. The difference is, Plantic dissolves when you put water on it.

But Plantic's disappearing trick is just one of its range of environmental benefits. Its primary ingredient (around 90 per cent) is corn, an annually renewable resource, unlike

the crude oil that goes into manufacturing traditional petrochemical plastic.

Fifty per cent less energy is used to manufacture Plantic, and the process does not use or create toxic compounds that can be associated with traditional plastic production.

After use, Plantic can be easily composted down to carbon dioxide and water, the basic ingredients for plants to grow more starch. The perfect circle.

In fact, one of the uses for Plantic is for seed and seedling trays, which melt into the ground they're planted in.

Most of the plastic that Plantic is replacing is never recycled and ends up in landfill. Even when Plantic is disposed of in landfill, because it dissolves in water it saves a substantial amount of space. Companies such as Cadbury Schweppes, Nestlé and Lindt & Sprüngli are already using Plantic, recognizing the environmental message it communicates to their customers.



**Garbology** This is the name of the academic study of refuse and trash. It began as a class project by two students at the University of Arizona, and the Garbology project itself started in 1971, under the direction of William L. Rathje.

Garbology and archeology often overlap. Ancient rubbish such as food remains and pollen traces of contemporary plants can be a unique source of knowledge. Garbology has unearthed very modern information, too. For example, it has revealed that the rate of natural biodegradation in landfills is much slower than expected.

Garbology can also mean an act of corporate espionage. Also known as 'dumpster diving', this can mean physically sifting through papers in bins or analyzing files in a computer's recycle bin.

# Spinach. It saved Popeye, now it could save the world.



Mobile communication and portable computing devices call for a continuous, self-sustained energy source, but present technology lags behind this need.

However, PhDs Shuguang Zhang and Marco Baldo at the Massachusetts Institute of Technology are collaborating with others on an idea that could solve this problem and ultimately revolutionize the way we generate energy on Earth.

Their Bio-Solar Energy Nanodevices go back to nature, to the process of photosynthesis. In the simplest terms, they are solar cells that use plant protein, spinach (although other green plant and algae material is viable) to convert sunlight directly into electrical energy.

Five trillion nanodevices can be packed on to a single square centimetre surface.



Almost all energy on Earth is obtained from photosynthesis through trees, green plants, and photosynthetic bacteria and algae. The fossil fuel and coal we use today comes from solar energy stored in them millions of years ago.

There are two plant photosynthesis systems, Photosystem I and Photosystem II, and they are the most efficient energy-harvesting systems there are. PS-II also produces hydrogen from water at the same time. Harnessing the energy and hydrogen these systems produce could provide us with clean and near-inexhaustible energy.

How does the Bio-Solar Energy Nanodevice work? As a first step, the developers used the simpler PS-I system.

The raw material is available at any supermarket of course, and it's purified to create the PS-I. The individual PS-I nanodevice is only around 20 nanometers in size, so five trillion can be packed on to a single square centimetre, two-dimensional surface area.

The purified spinach (the protein) has to be stabilized on to gold surfaces. This is achieved by using novel molecular-designed peptide detergents discovered by Shuguang Zhang. They act as artificial membranes to stabilize the finicky protein complexes, and they effectively preserve the integrity of the PS-I complex in dry form for over 500 hours (three weeks).

The peptide detergent PS-I complex system produces not only electric current but also electric voltage.

By definition, the components of nanotechnology are 100,000 times thinner than a human hair, but these Bio-Solar Energy Nanodevices could be the start of something very big in the harvesting of energy.



**Peptides? Detergents?** There's a basic set of molecular building blocks in nature. This includes 20 amino acids, a few nucleotides (the structural units of nuclear acids such as RNA and DNA), around a dozen lipid molecules, and two dozen sugars.

Biotechnologists have learned how to manipulate these building blocks. Using amino acids, for example, to create peptides.

These peptides are molecular architectural units. In water and in body fluids, they form well-ordered nanofibre scaffolds useful for growing three-dimensional tissue and for regenerative medicine.

All biomolecules, including peptides, naturally interact and self-organize to form well-defined structures.

The peptide detergents discovered by Shuguang Zhang probably work (in a similar way to other detergents) by surrounding parts of the spinach in the nanodevice and preventing them from interacting with water molecules. By doing this, the detergents maintain the overall structural integrity of the PS-I complex and retain its biological function.

# A FEAST OF IDEAS

The first Saatchi & Saatchi Award for Innovation in Communication was presented at the State of the World Forum in San Francisco on 29th October 1998.

The night before, we'd hosted a dinner for the finalists at a local restaurant.

The guests arrived quite reserved, and somewhat nervous.

(Particularly Ji Lee, a young designer at Parsons School of Design in NY, who had suddenly found himself sitting next to Buzz Aldrin.)

Many of the finalists were curious as to exactly why we were doing this award.

We explained Saatchi & Saatchi's inspirational dream 'To be revered as a hothouse for world changing ideas', and the part the award would play in helping to bring both their dreams and ours to life.

The purpose was to bring their innovations out of the dark into the light on a world stage.

(The finalists were being profiled on the CNN website for the whole week, plus a plethora of other media interviews)

Then there was the prize itself. Worth U.S $100,000. The largest amount at that time for any award of this nature.

The body language around the table was interesting.

At first everyone sitting bolt upright, very restrained.

As we talked about the thinking behind the award and the possibilities it might bring, people started to become very engaged.

No longer sitting back in their chairs, they were leaning forward, on their elbows, into the table.

And then the most fantastic thing happened.

Some had already been aware of the work of other finalists and they started to swap their expertise.

"I really like your concept but it could be much lighter without that battery.

I can help you with that." etc; and business cards started flowing.

It had turned into a hothouse for world changing ideas.

BOB ISHERWOOD

2

# The Language World

# Read a foreign language you've never seen before.

Quicktionary
FINALIST, 1998

A Quicktionary looks like a big pen. When you run the nib of the pen over a piece of foreign text, it scans the words and immediately comes up with a translation on its built-in screen.

Quicktionary started with a question.

In the mid-1990s, an Israeli student, Adi Lipman, asked his entrepreneur father Aharon why a device didn't exist that would instantly translate words on the printed page.

In response, Aharon observed that flatbed scanners existed and so did electronic dictionaries, but no one had combined the two into something portable.

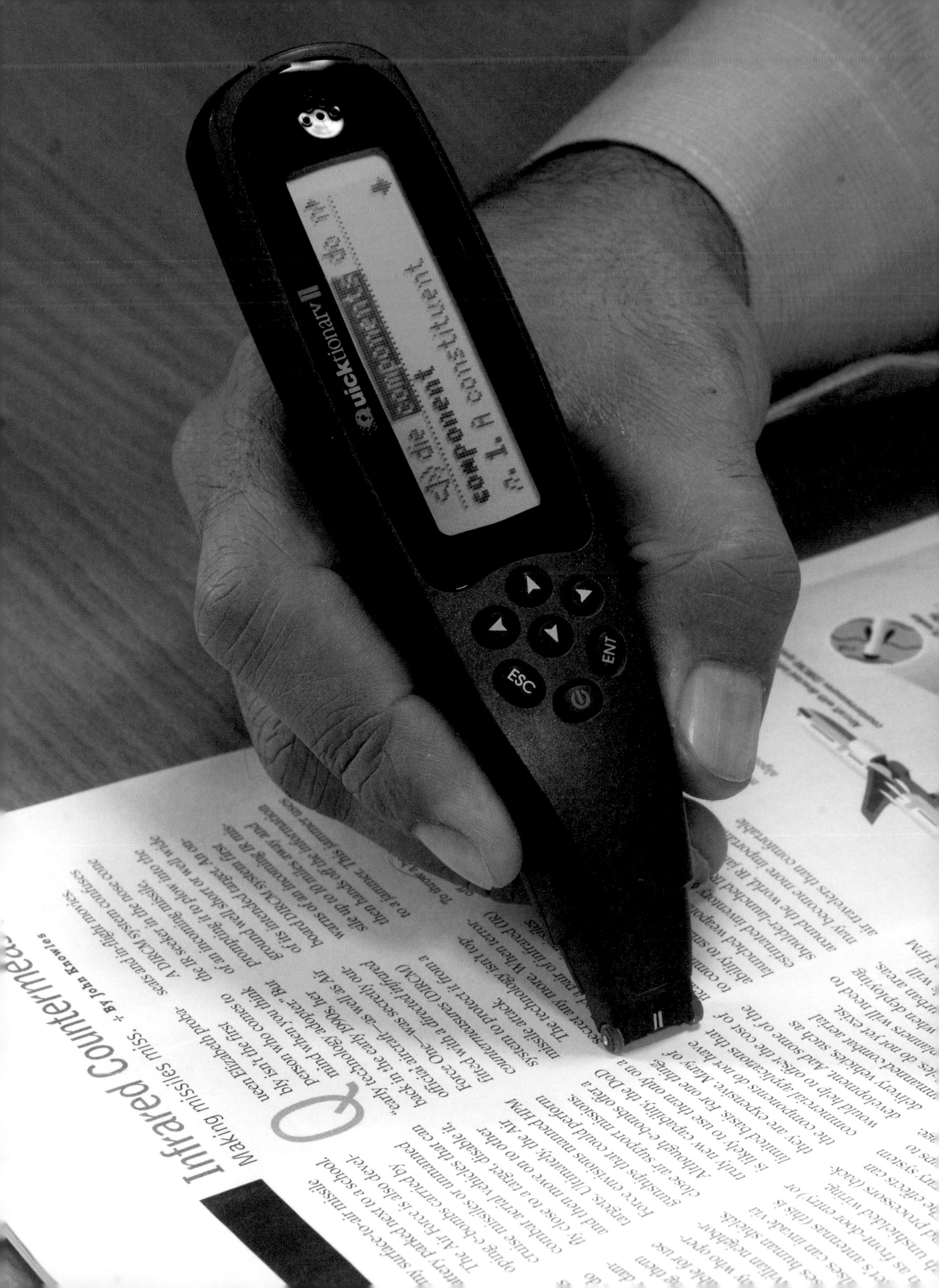

quicktionary II
component
1. A constituent
ESC
ENT

Aharon decided to do just that, and the first product was launched in 1997.

Since then, the range of Pens produced by WizCom Technologies Ltd has grown significantly.

There are Pens now that not only translate and define individual words, they also provide translations and explanations of words in context, such as phrasal verbs and idioms. This allows for a better understanding and interpretation of the source language.

The Pens also use a unique Text-To-Speech function that enables users to actually listen to the words or sentences they have scanned.



The developers at WizCom recognize that every language has different forms and structures, and they adapt the pens accordingly. For example, German Pens can read split verbs, a grammatical form unique to the language. Vertical scanning has also been created for Pens supporting Chinese text.

The Pens are not solely translating devices either. Some are able to store text and transmit it to computers and

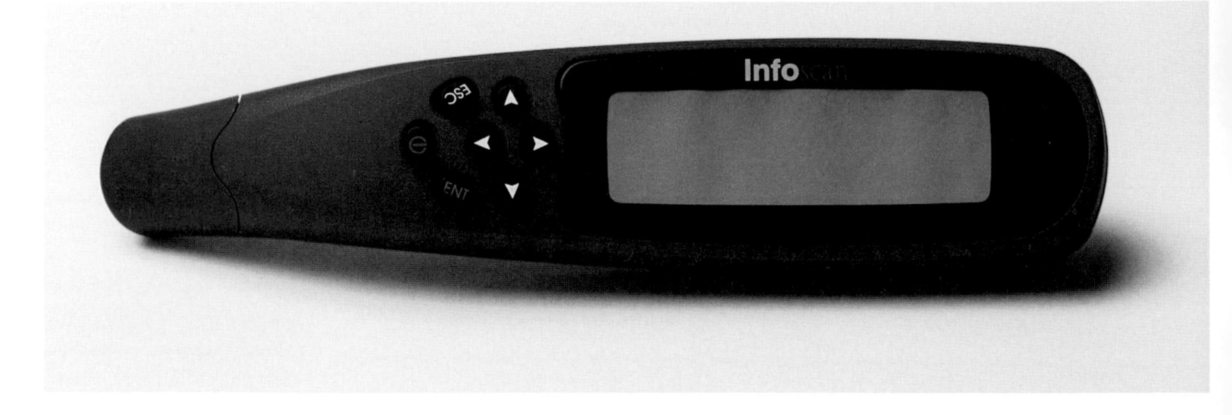

handheld devices, via infrared communication or by direct connection to a PC. Text can also be scanned directly to a PC and transferred at a later time.

There is even a Reading Pen®, which displays words in large type sizes, divides them into syllables and letters, and plays the sound of the words to help people with reading difficulties such as dyslexia.

Some of WizCom's pens have built-in licensed dictionaries from publishers such as Houghton Miffin and Oxford University Press, while others feature the company's own 30 proprietary databases. In all, WizCom has 25 language products and is the clear market leader.

With English alone spoken as a second language by around 320 million people, Quicktionary's potential market is vast. And so is its potential for making communication possible between different communities, and communication clear between people sharing the same language.

**The language league** The human population of the Earth is around six billion with about 7,000 languages between them.

Estimates of how many people speak a language vary considerably. For example, estimates for English range from 275 to 550 million and for Hindi from 150 to 350 million. Compare these two leagues:

*From infoplease.com*
(sources: Ethnologue 13th edition and others):

1. Mandarin Chinese (1,075,000,000)
2. English (514,000,000)
3. Hindustani (496,000,000)
4. Spanish (425,000,000)
5. Russian (275,000,000)
6. Arabic (256,000,000)
7. Bengali (215,000,000)
8. Portuguese (194,000,000)
9. Malay-Indonesian (176,000,000)
10. French (129,000,000)

*From geography.about.com*
(source: 2005 CIA World Factbook):

1. Mandarin Chinese (882,475,389)
2. Spanish (325,529,636)
3. English (311,992,760)
4. Hindi (181,780,905)
5. Portuguese (178,557,840)
6. Bengali (172,756,322)
7. Russian (146,327,183)
8. Japanese (128,278,015)
9. German (96,047,358)
10. Wu Chinese (77,998,190)

Please note: the CIA World Factbook estimates the world's population at 6,446,131,400 and expresses numbers of speakers as percentages of this.

But what happened to Arabic?

Quicktionary
FINALIST, 1998

# Say what you think without speaking.



This is the story, so far, of an innovation that could have come straight from science fiction.

Dr Chuck Jorgensen, at NASA's Ames Research Center, is working on a new form of human communication – Subvocal Speech Recognition.

Instead of using sound to communicate, Subvocal Speech uses the tiny neural impulses in the human vocal tract.

These neural impulses are called electromyographic (EMG) signals and they arise from commands sent by the brain's speech centre to tongue and larynx muscles, to be turned into audible words.

Subvocal Speech intercepts EMG signals before any sound is made, using sensors placed on the 'speaker's' neck, and infers what words would have been said. So whether or not recognizable sounds are made is no longer important. What's more, ambient noise or even intelligibility problems, such as accents, no longer hamper communication.

Key is that the neural signals produced are consistent, and consistency relies on a common language between speaker and listener.

Separating speech understanding from sound generation will give rise to significant new communication options.

For example, handicapped people with some consistent voice-muscle behaviour may now be understood when before, verbally, they couldn't. Subvocal Speech can also work for laryngectomy patients whose vocal cords have been removed, and for the deaf who cannot hear what their speech sounds like to others.

Subvocal Speech can overcome major speech-generation and intelligibility problems related to underwater diver mouthpieces, pressurized fire suits, different breathing-gas mixes and extreme external noise.

Another benefit is greater privacy. If sound isn't produced, it can't be overheard, which has implications for both confidential business conversations, and for personal security, when passwords need no longer be typed or spoken in public places.

Dr Chuck Jorgensen controls a web browser subvocally. At bottom left are two subvocal signals coming from the neck electrodes from which an alphabetic character is deduced (middle left side 'A').

Jorgensen guides, subvocally, a graphic simulation of a Mars rover in real time. Electrodes on his neck pick up the signals from his nervous system to control the graphic.

A silent mobile or cell phone might be a revolutionary telecommunication spin-off, and the laboratory at the Ames Research Center has demonstrated a first step towards this.

In a similar way, silent communication between humans and machines is possible.

Subvocal Speech has reached a point similar to the early stages of auditory speech recognition. But silence could soon be making a very big noise.



**What you don't need to think before you speak** There's a lot you don't have to consider when you speak, because your brain does it for you.

It can take your brain half a second or more to prepare for you to speak a sentence, because of the processes involved.

As well as selecting the words, your brain also works out the correct emotional emphasis, and it plans any facial or hand gestures to go with the message.

Your brain even calculates how much breath your lungs will need to deliver the planned sentence, and instructs them accordingly.

# 3-D alphabets. An A to Z.

Univers Revolved
FINALIST, 1998

Univers is the name of a plain, widely used typeface.

Univers Revolved is rather different. It's a three-dimensional alphabet, which began as an experimental typography project during a computer class.

The student was Ji Lee, who was born in Seoul, South Korea, moved with his family to São Paulo, Brazil when he was 10, and headed to New York in 1991 to study arts and design at the prestigious Parsons School of Design.

Univers Revolved transforms two-dimensional reading into a three-dimensional experience.

'The mysterious shapes of the letters turn reading into an imaginative mind game. They add an intriguing twist to visual communication,' says Ji Lee.

'Unlike the Latin alphabet, which contains both symmetrical and asymmetrical letters, those of Univers Revolved are all bilaterally symmetrical, so it opens up the possibilities for bilateral reading.'

His 3-D letters can be stacked, placed in a circle and be read in any direction, or set in motion. Equally they could be turned into pieces of furniture, building blocks or chocolate candies. Ji Lee stresses how his invention helps us to rethink communication. 'Univers Revolved is a pause from our linear way of thinking and presents reading as multi-dimensional and fun. It invites readers to think outside the conventions and find joy in their new discoveries.'

Ji Lee has published a visual puzzle book about his invention, called *Univers Revolved, a Three-Dimensional Alphabet*. It won the New York Book Award and it's in bookshops worldwide – the only book written entirely in Univers Revolved, surrounded, no doubt, by many other books typeset in traditional Univers typeface.

You can read all about Ji Lee's typeface by visiting www.universrevolved.com.

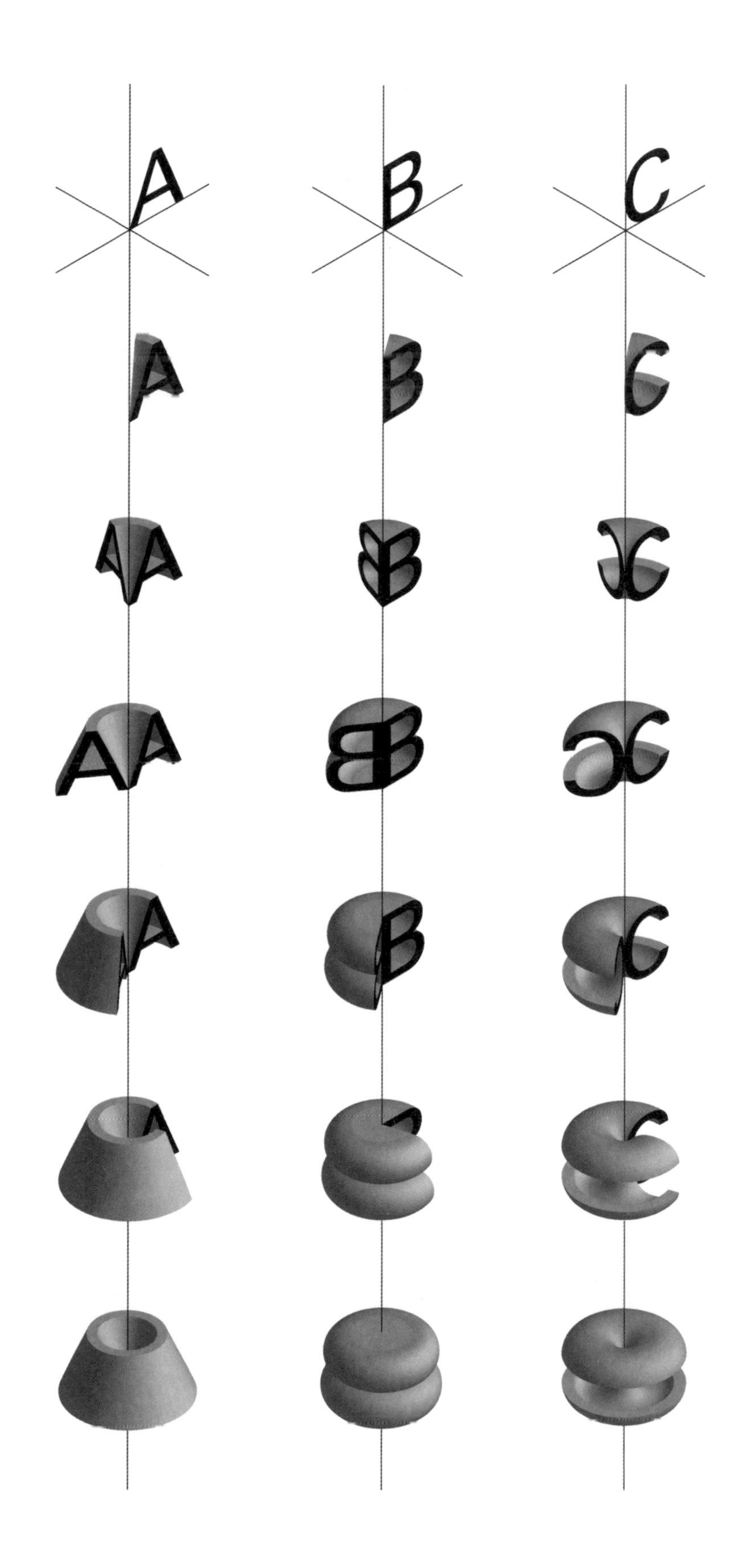

Univers Revolved
FINALIST, 1998

**Alphabets** We've been on Earth for millions of years, but we've only been writing for the last 6,000 of them.

The oldest known examples date from the fourth millennium BC and were found on clay tablets in Mesopotamia (the land between the Tigris and Euphrates rivers).

The first alphabet only emerged around 1,000 BC. But it revolutionized writing by making it much simpler than the hundreds of symbols and hieroglyphs that had to be learned by Chinese and Egyptian writers, for example.

The Phoenicians are credited with creating this first alphabet, and since they were principally sailors and merchants their alphabet spread with their travels.

Our own Latin alphabet emerged around the third century BC, at first consisting of 19 letters.

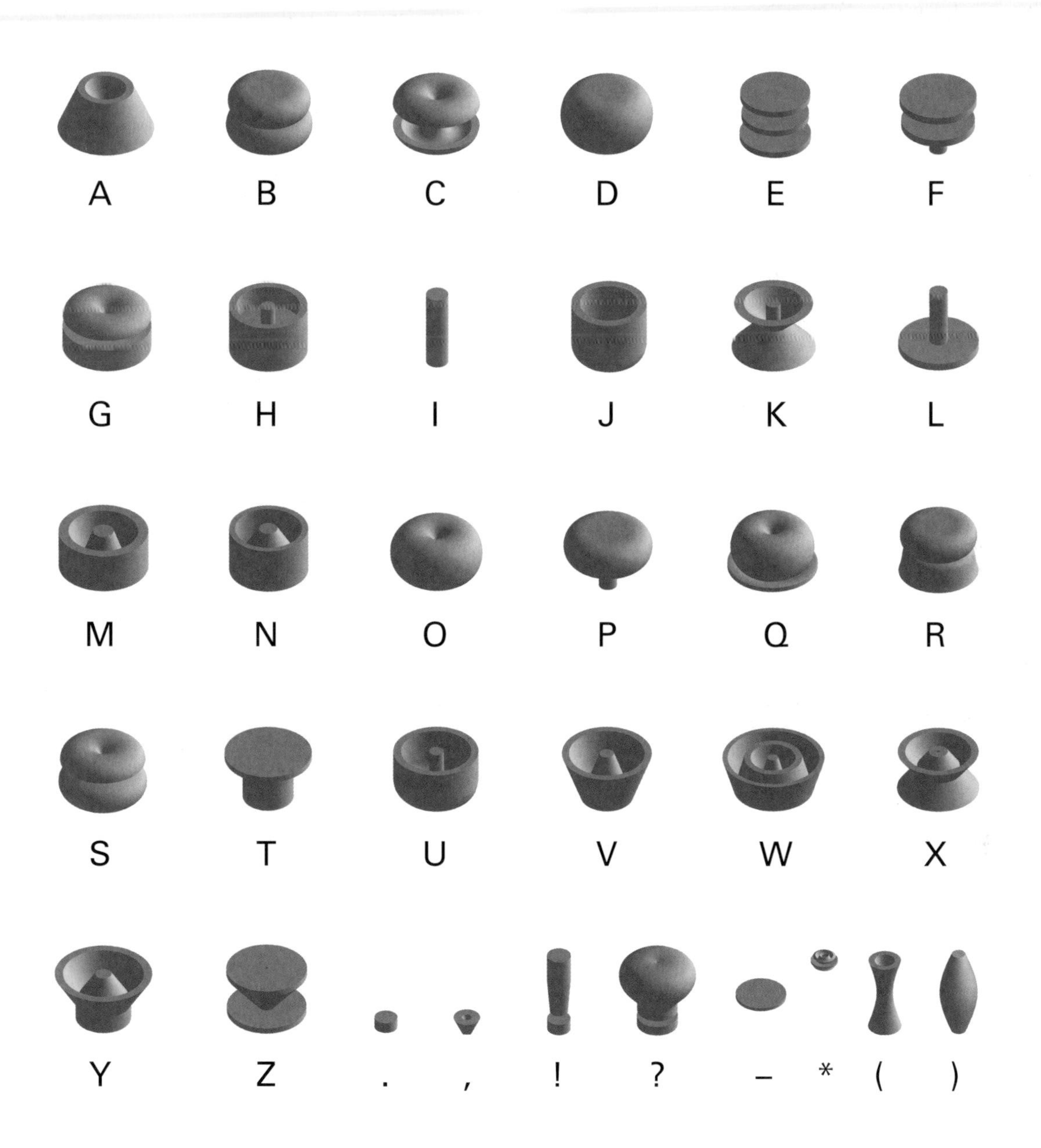


Full character set of Ji Lee's Univers Revolved, with corresponding original Univers characters by Adrian Frutiger.

# JUDGES. THE FIRST PARTICIPANTS.

From the outset we knew the success of the award would rest on the quality of the entries.

To help attract the world's best innovations we decided to put together a list of judges representing some of the world's foremost thinkers and pioneers.

Could we get them?

That would be the first internal test of the Award concept. After an initial letter of approach (penned by Richard) some invitees wanted more information and personal follow up.

That fell to me.

It turned out to be a time for a number of personal heroes to become personal friends. Picking up the phone and calling people like Buzz Aldrin and the late Tibor Kalman was a thrill.

The thought of meeting Edward de Bono, long recognized as the foremost international authority on conceptual thinking, was a little daunting. In the event, it was totally inspiring.

Dr de Bono had expressed an interest in judging but there was something he wanted to discuss.

Were we willing to include a special award for the idea that met de Bono's criteria of 'simple, practical, effective, and in use?' We agreed. As for the physical award itself de Bono had his own idea.

"It should be made from glass, glass is made from sand and sand has no intrinsic value."

We had the first 'De Bono Medal' as we'd called it made by a small Italian glass blowing firm in Murano, Italy.

It had been Richard's idea and by sheer co-incidence it turned out they were personal friends of de Bono's.

Over the years Edward and I have become good friends too, and it's somewhat amusing now to watch others approach him with trepidation.

My first meeting with Edward had been in Sydney and also by co-incidence we found ourselves flying back to the UK together.

The pilot on his 'meet and greet' tour clearly recognized de Bono from appearances on Australian TV. Happily chatting to every other passenger he warily passed around us and the challenging mind game Edward was demonstrating on his tray table.

3

# The Knowledge World

# What do you know? An encyclopedia that's up to date.

Wikipedia
FINALIST, 2005

Wikipedia is a multi-lingual, web-based, free-content encyclopedia founded by Jimmy Wales and named by Larry Sanger. Jimmy says Wikipedia is 'an effort to create and distribute a free encyclopedia of the highest possible quality to every single person on the planet in their own language'.

The Wikipedia site went live on 15 January 2001 and now, so far, Wikipedia contains more than 3.5 million articles in over 100 languages.

The articles are written by volunteers, and it's described as 'the largest collaborative effort of its kind to bring people of all ages and backgrounds together to create a compendium of human knowledge'. The project uses wiki software. Although Wales and Sanger dispute whose idea it was to do so, what it means is that articles can be added or changed by almost anyone, and supports the belief that collaboration among users will improve articles over time. There are two clear manifestations of this.

One is that short 'stub' articles, containing little more than a definition, are quickly expanded by teams of dedicated editors who add information, refine language, present opposing ideas, insert maps and illustrations, and include hyperlinks to related articles. The second is that, because articles are always subject to editing, Wikipedia does not declare any article finished.

By its nature, Wikipedia's content is constantly revised, so news appears as news rather than as historical reflections as it might in a traditional encyclopedia.

Although Wikipedia requires contributors to observe a 'neutral point of view' when writing, and to not include original research, they readily acknowledge on the site that they have their critics.

One criticism they cite is that allowing anyone to edit makes Wikipedia an unreliable work. They also quote *Encyclopædia Britannica*'s executive editor, Ted Pappas. 'The premise of Wikipedia is that continuous improvement will lead to perfection. That premise is completely unproven.'

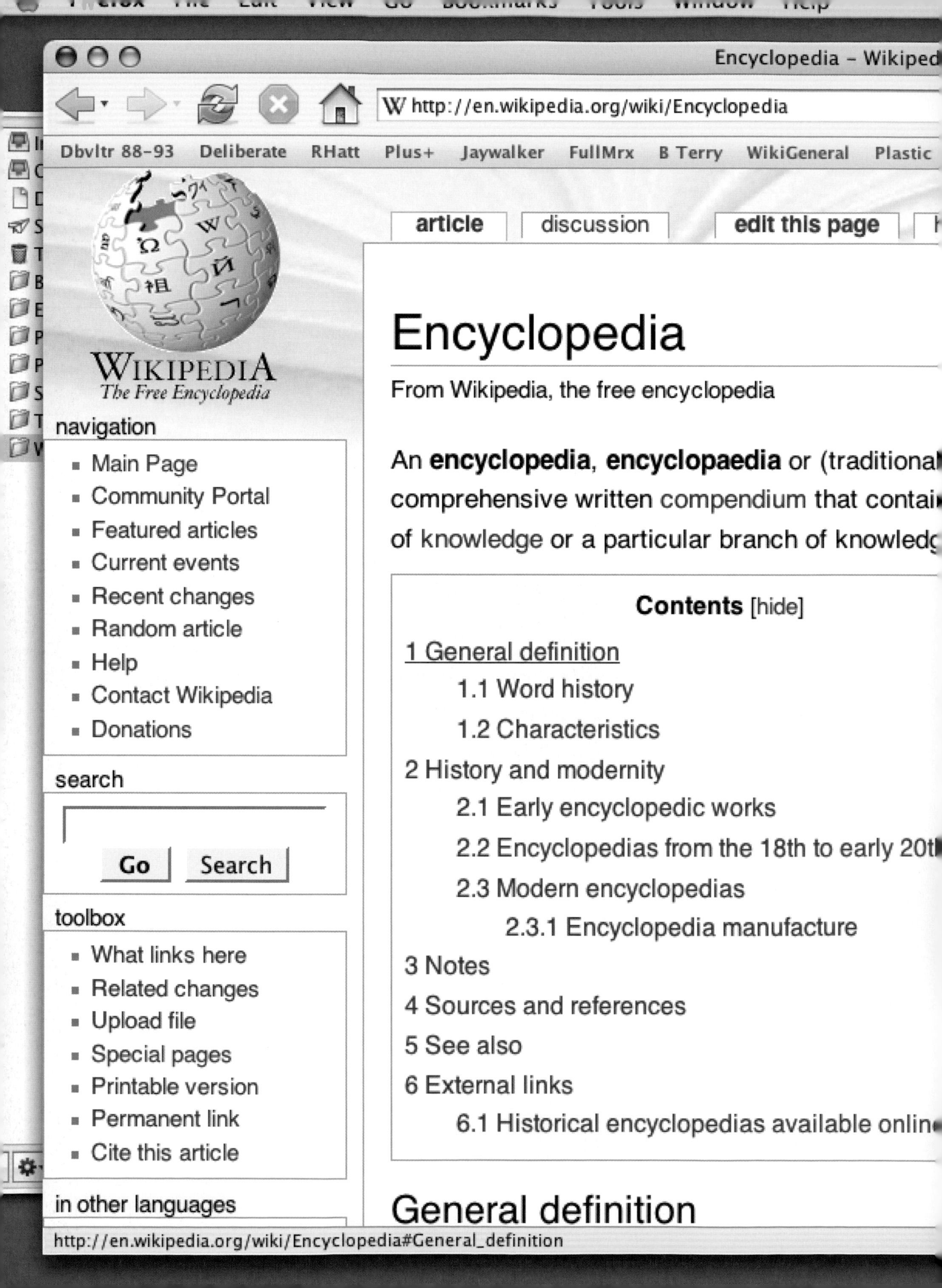

Encyclopedia - Wikiped
http://en.wikipedia.org/wiki/Encyclopedia
Dbvltr 88-93
Deliberate
RHatt
Plus+
Jaywalker
FullMrx
B Terry
WikiGeneral
Plastic
WIKIPEDIA
The Free Encyclopedia
article
discussion
edit this page
Encyclopedia
From Wikipedia, the free encyclopedia
An encyclopedia, encyclopaedia or (traditiona
comprehensive written compendium that contai
of knowledge or a particular branch of knowled
navigation
Main Page
Community Portal
Featured articles
Current events
Recent changes
Random article
Help
Contact Wikipedia
Donations
Contents [hide]
1 General definition
1.1 Word history
1.2 Characteristics
2 History and modernity
2.1 Early encyclopedic works
2.2 Encyclopedias from the 18th to early 20t
2.3 Modern encyclopedias
2.3.1 Encyclopedia manufacture
3 Notes
4 Sources and references
5 See also
6 External links
6.1 Historical encyclopedias available onlin
search
Go
Search
toolbox
What links here
Related changes
Upload file
Special pages
Printable version
Permanent link
Cite this article
in other languages
General definition
http://en.wikipedia.org/wiki/Encyclopedia#General_definition

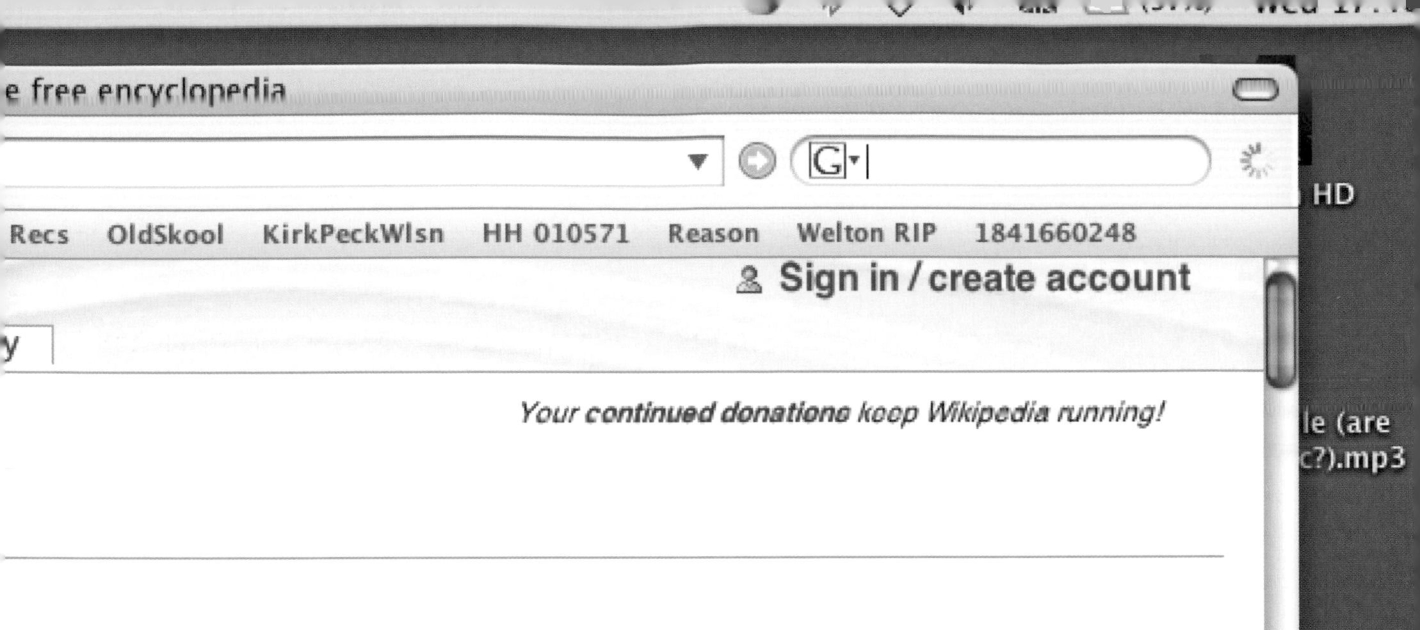


**ncyclopædia,**[1] is a
formation on all branches

Brockhaus Konversations-Lexikon, 1902

tury

[edit]

The fact remains that Wikipedia has been used by news media and academics, in books, conferences and even court cases. The parliament of Canada website refers to Wikipedia's article on same-sex marriage in the 'further reading' list for Bill C-38. With an average of 5.45 million visits per hour on all its languages, Wikipedia is currently among the 50 most-visited websites in the world. It's clearly satisfying a great hunger for knowledge.

**Where Wikipedia gets its name** Wikipedia uses wiki software. 'Wiki' is a term originally used for the Wikiwikiweb.

'Wikiwiki' is the name of the shuttlebus at Honolulu International Airport and it comes from the Hawaiian word 'wiki', meaning 'quick'.

Incidentally, just 2,000 people are native Hawaiian speakers. But at least now, thanks to Wikipedia, millions of people are aware of one Hawaiian word.

# What's the word for 'saved' in every human language?

Rosetta Disk
FINALIST, 2000

It's been estimated that we lose one language every week. This means that 50 to 90 per cent of the world's languages will disappear during this century.

There's little or no significant documentation for many of the 7,000 or so languages on Earth, and much of the work that has been done, especially on smaller languages, is hidden away in private research files or poorly preserved in underfunded archives.

Significantly, 96 per cent of the world's population uses only four per cent of its languages.

This threat to our legacy of linguistic diversity prompted the establishment of The Rosetta Project by The Long Now

Each Rosetta Disk is housed in this four-inch sphere. Upper hemisphere is optical glass and doubles as a 6x viewer. Lower hemisphere is stainless steel.

Disk is etched with a central image of Earth and, in eight major languages, the message: 'Languages of the world: This is an archive of over 1,000 human languages assembled in the year 02002 C.E. Magnify 1,000 times to find over 15,000 pages of language documentation.'

Foundation. The Project is a global collaboration of language specialists and native speakers who are working to build a publicly available archive of all documented human languages.

The Project's ambition is to create a reference work that's relevant for academic researchers and educators, as well as for native communities who need materials to help with language-revitalization work.

There are three forms for the archive: online, a monumental single-volume reference book, and the highly innovative Rosetta Disk. It's an extreme-longevity micro-etched three-inch nickel disk, which its creators describe as a contemporary 'Rosetta Stone'.

The archive and, in turn, the information etched on the disk, is an expansion of the parallel text structure of the Rosetta Stone, which, by cross-referencing its triple inscriptions allowed Egyptian hieroglyphs to be deciphered for the first time.

The disk is etched with 10 descriptive components for 1,000 of the languages embraced by the Project, ranging from maps of each language's geographical distribution, to a parallel translation of Genesis Chapters 1–3 in all 1,000 languages. (Biblical texts are the most widely and carefully translated writings on the planet.)

Because the encoding on the disk is analogue, there is no platform or format dependency, guaranteeing readability despite changes in digital systems.

The etched information on the disk is read using a 1,000x optical microscope. This level of readability gives the disk a storage capacity of around 15,000 pages of text.

The intention is to mass-produce Rosetta Disks (reflecting the archive principle of 'lots of copies keep stuff safe') and distribute them to thousands of interested individuals, organizations and native communities. The possibility is that the Rosetta Disk, like its predecessor, may serve as a key to past languages for archeologists some time in the deep future.

Appropriately, a copy of the Rosetta Disk was on board when the European Space Agency Rosetta mission was launched in February 2004. The role of the mission is to unlock the mysteries of our Solar System's oldest building blocks, the comets.



**The Rosetta Stone** The Rosetta Stone is a slab of dark-grey granite-like stone with a pink vein, found near the town of Rosetta, now Rashid, in Egypt in 1799.

The inscribed text on the Stone's surface dates from 196 BC and is a decree passed by a council of priests affirming the royal cult of the 13-year-old Ptolemy the Fifth on the first anniversary of his coronation.

The text appears in three forms: in Egyptian hieroglyphic and Egyptian demotic scripts, and in Greek. Demotic is a cursive script, a more easily written form of hieroglyphic.

Because Greek was readily understood, the parallel texts allowed Egyptian hieroglyphs to be deciphered for the first time.

The Stone was discovered by an officer in Napoleon's army, but was captured by the British two years later and placed in the British Museum in London.

The work of deciphering the hieroglyphics was undertaken principally by the French scholar Jean-François Champollion in 1822, and by the English physician and physicist Thomas Young.

# ADVISING A VISIONARY

The first recipient of the De Bono Medal was Professor Joshua Silver (Self-Adjustable Spectacles.)

He said at the time that it was worth more to him than the money prize.

He was a big de Bono fan and would have loved to have met him.

(Edward was unable to make the presentation)

We were able to put this together a year later at a wonderful dinner Professor Silver hosted in his rooms at Oxford.

Other guests were Kevin Roberts, Bob Seelert our Chairman and his wife Sarah (another de Bono fan), Mary Ellen Barton (manager to Edward) and the New Zealand entrepreneur Sir Michael Fay (the financier behind New Zealand's successful America's Cup challenge.)

Professor Silver asked us for advice on how, faced with a complexity of different business options, he could best move his project on.

Sir Michael offered "Josh, my advice is, it's time to fire yourself."

Subsequently Josh did get a business manager but the thought of an inventor firing himself from his invention was pretty novel to me.

4

# The Technology World

# The new plane that looks like an old lawnmower.



FanWing is a revolutionary aircraft, both technologically and in the way it looks. It's been described as a 'conspicuously odd bird' and its inventor, Patrick Peebles, reported that its appearance at the UK's Farnborough Air Show prompted a lot of people to ask him if it would also mow their grass.

What makes FanWing different is the cross-flow fan along the leading edge of each wing. The fan pulls the air in at the front and accelerates it over the trailing edge of the wing. By transferring the work of the engine to the rotor, which spans the whole wing, the FanWing accelerates a large volume of air and achieves a high lift-efficiency.

FanWing
EDWARD DE BONO MEDAL WINNER, 2002

This translates as a lot of lift for very little power. The lift generated is three times that achieved by a helicopter for the same power used or fuel consumed.

So where traditional machine-powered flying is notoriously fuel-guzzling, and high emission-generating, FanWing isn't.

The benefits of FanWing go beyond the environmental. Short take-off and landing (with the distinct possibility of vertical take-off). Reduced noise. Stable flight because the aircraft is not sensitive to the angle of incoming air. It won't stall. It's simple and economical to construct. It's highly manoeuvrable and easy to fly.

Patrick Peebles sees a number of uses for FanWing, including short-range freight delivery, as an 'Air-Truck'. There's also a wide range of unmanned uses for FanWing. As a UAV (unmanned aerial vehicle), FanWing could have both military and civilian applications, including surveillance, minesweeping, archeology, traffic control, irrigation, pest control, fire watch and control, and ecological monitoring.



**How normal planes fly** All flight – mechanical or natural – needs wings. As a wing cuts through the air, air flowing around it creates lift, pushing the wing upwards.

Conventional aeroplane wings have an arched cross-section called an aerofoil. The top of the wing is curved, the bottom is flat.

Air moves rapidly up and over the curved top of the wing. Because it moves faster, its pressure drops. By contrast, the air passing over the flat underside of the wing moves more slowly so its pressure is higher, forcing the wing upwards.

So, to take off, a plane speeds along the runway until it is going so fast that the lift generated under its wings overcomes its weight and it flies.

To climb, the plane's nose is tilted upwards so the angle at which the wing meets the airflow increases, producing more lift. But if the wing meets the incoming air too steeply, the plane stalls, lift is no longer generated and the plane falls.

# 3-D images that save lives.

Crossed Beam Display
FINALIST, 1993

Although we exist in a three-dimensional world and get 80 per cent of our sensory input through our eyes, much of the information we receive comes in two-dimensional form. The funny glasses we wear in cinemas simply trick us into thinking we're seeing a 3-D image. In reality, it's still only 2-D, because the cinema screen is only ever flat.

To create genuine 3-D imagery, Dr Elizabeth Downing treated some glass with an active ion, then intersected two infrared laser beams through it, which generated visible light. By scanning the lasers around, real 3-D images can be created that have volume and depth. Dr Downing's Crossed Beam Display was born.

Users can look into an image chamber and view objects from all directions – front, back, sides, top.

The first Display built by her company, 3-D Technology Laboratories, was delivered to NASA's Goddard Space Flight Center in 1998.

Of the many applications for the Crossed Beam Display, a number of them could be life-saving.

Imagine a team of surgeons planning to remove a tumour from the brain of an epileptic patient. The patient's brain has been scanned using MRI, and the information, the volumetric data, is displayed inside a real 3-D image chamber. The surgeons position themselves around the Display and plan the procedure using their own interactive 3-D cursors. It means that doctors can explore inside the human body without making a single incision.

At a busy airport, air-traffic controllers could monitor flights by tracking miniature planes flying around inside a scaled-down three-dimensional model of their airspace. The information source would be the airport's radar. Potentially hazardous situations would be signalled by making any planes flash that are too close to each other.

Crossed Beam Display
FINALIST, 1998

**The 3-D illusion** 3-D, or stereoscopic photography, creates paired images that imitate the way a human's two eyes see the world.

Our eyes are about 65 mm (2.5 in) apart, and so are the lenses of the two cameras that capture the images.

Two projectors are used to show the images. These are each fitted with a polarizing filter fixed at right angles to the other. The spectacles that viewers wear have corresponding polarizing filters so that each eye sees only the image intended for that eye. Similarly, 3-D film is also based on recording right-eye and left-eye views. These are either created on separate strips or side by side on one wide film.

The process mimics our binocular vision and fools us into thinking something has depth when in reality it is flat.

# Make your fingers smaller than they are.

Quadkey
FINALIST, 1998

David Levy sums up his approach to inventing as 'I go looking for trouble'. This includes future problems too, and well ahead of time Levy saw technology getting more complex as, at the same time, gadgets were getting smaller.

Miniaturization and only a 12-button keypad to work with seemed like the kind of trouble Levy looks for.

You can't make human fingers smaller so, instead, Levy redefined the button. Originally called Quadkey, Fastap technology uses raised and lowered keys, together with error-prevention software, to allow up to three times as many keys in a product without compromising on ergonomics.

First Fastap concept model.

Quadkey
FINALIST, 1998

Commercially available Fastap LG AX490.

The result is tremendous extra functionality from keypads that work flawlessly with any hand size.

He filed patents, and waited for the world to catch up. In November 2004, Telus Mobility, a mobile operator in Canada, started selling its first Fastap phone, manufactured by LG and called the 6190.

The phone sold extremely well and, just as important, Telus discovered that the interface changes consumer behaviour. Specifically, the increased usability of Fastap roughly doubles data usage, with accompanying revenue implications for the operator.

In the third quarter of 2006, another Fastap phone was launched in Canada and the United States.

The technology has also expanded from being an alphanumeric solution to enhancing the standard 12-button keypad with extra keys that provide one-touch access to MP3, web services and features otherwise buried deep within the phone.

Fastap technology is also being applied to enhancing QWERTY phones, both by making tiny versions easier to use and by adding large numbers of extra keys as they are needed.

**Making people smaller** The miniaturizing of technology is accelerating. Nanotechnology is a reality. But it's only in science fiction that people have been miniaturized. Perhaps the best-known example is the 1966 film *Fantastic Voyage*, in which a scientist, Jan Benes, working behind the Iron Curtain, figures out how to shrink matter for an unlimited time to overcome the 60-minute maximum then possible. With the help of the CIA, Benes escapes to the West, but he's left in a coma with a blood clot on his brain after an assassination attempt.

The US government wants to save his life – and his secret of unlimited miniaturization.

A group of scientists board a submarine, the *Proteus*, which is then miniaturized and injected into Benes.

The scientists have just 60 minutes to find and repair the blood clot, before returning to normal size....

Quadkey
FINALIST, 1998

# No more spaghetti or jam in phones.

Sprint Integrated On-Demand Network
FINALIST, 1998

The Sprint Integrated On-Demand Network set out to solve a growing problem in the relative technological dark ages of the 1990s.

At the time, households and businesses were facing a proliferation of telecom offerings. Phones, faxes, video (including videoconferencing) and Internet access – all of which were being offered by a plethora of operators, and all of which competed for the limited bandwidth available. The Sprint ION was conceived to provide all telecom needs, on a real-time basis, down just one phone line.

Sprint headquarters with diagram of Integrated On-Demand Network superimposed.

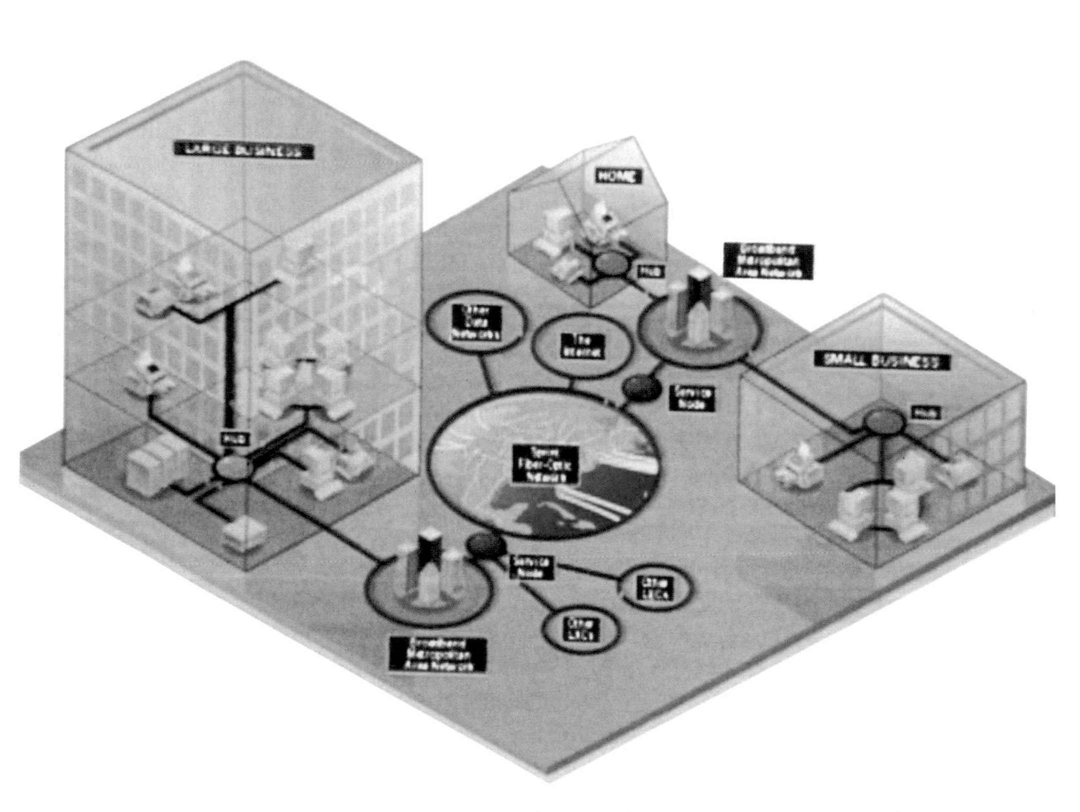


The benefits of Sprint Integrated On-Demand Network covered large and small businesses and the home (including employees working at home).

The idea revolved around a hub with the software to understand the services the user was asking for at any given time.

ION was a new and much cheaper method of making all the communication systems work together in the simplest way possible, by enabling the available bandwidth to be shared between them efficiently.

ION did away with traditional switch centres and employed components with more relevance to the Internet age: high-speed switches, data-packet routes and fibre optics. Among other things, the new components would speed up Internet sessions by 100 times.

Sprint likened the difference between ION and its competitors to a multi-media computer and a roomful of pocket calculators.

Sprint's vision for ION was that it would become a low-cost standard for the telecom industry, one that would create unprecedented solutions for businesses that need communications that are both cost-efficient and high-powered.

It's hardly surprising that many of the patents developed for ION are what power broadband networks today. So the vision of ION is becoming a reality. It just isn't called ION.

**Broadband Internet access** Broadband Internet access is a high data transmission rate Internet connection.

DSL (digital subscriber line) and cable modem are popular consumer broadband technologies capable of transmitting 256 kilobits or more per second, which is around nine times the speed of a modem using a standard digital telephone line.

Broadband Internet access took off in the early 2000s and one study in the USA found that its usage grew from six per cent in June 2000 to over 30 per cent by 2003.

Providing broadband access in low-population areas is a challenge because each rural customer may need expensive equipment to get connected. This mirrors the issue with electricity a century ago in the USA. Urban areas had electric lighting as early as 1880, but remote rural areas were still without it until the 1940s.

Sprint Integrated On-Demand Network
FINALIST, 1998

# Walls can be windows and so can floors.



SmartSlab is a display system that's also a building material, so it allows all kinds of digital images and communication to become an integrated part of an environment.

SmartSlab is made up of slim 60 x 60-cm (24 x 24-in) tiles containing hexagonal pixels, which are seamlessly joined and sandwiched between two layers of polycarbonate sheet for extra strength.

This fascia is then connected to a local host computer (which can be a standard PC) as a source for displaying every and all forms of visual materials, from artwork to advertising information to whole web pages, including sound if required.

It's the world's toughest modular structural 'tile' and can be part of any wall, floor or ceiling, indoors or outside.

It's possible to 'wrap' an image around a building and it's the first-ever display system to measure 'dwell time' – how long people stand and look at it.

The inventor, Professor Tom Barker, used to run a product-design firm, B Consultants. The company was working on installations in London's Millennium Dome project.

Barker had been experimenting to produce a translucent building structure for the 'Mind Zone', designed by the architect Zaha Hadid. At the same time, he saw how exhibitors were having trouble with display technology and getting the lighting in the Dome right.



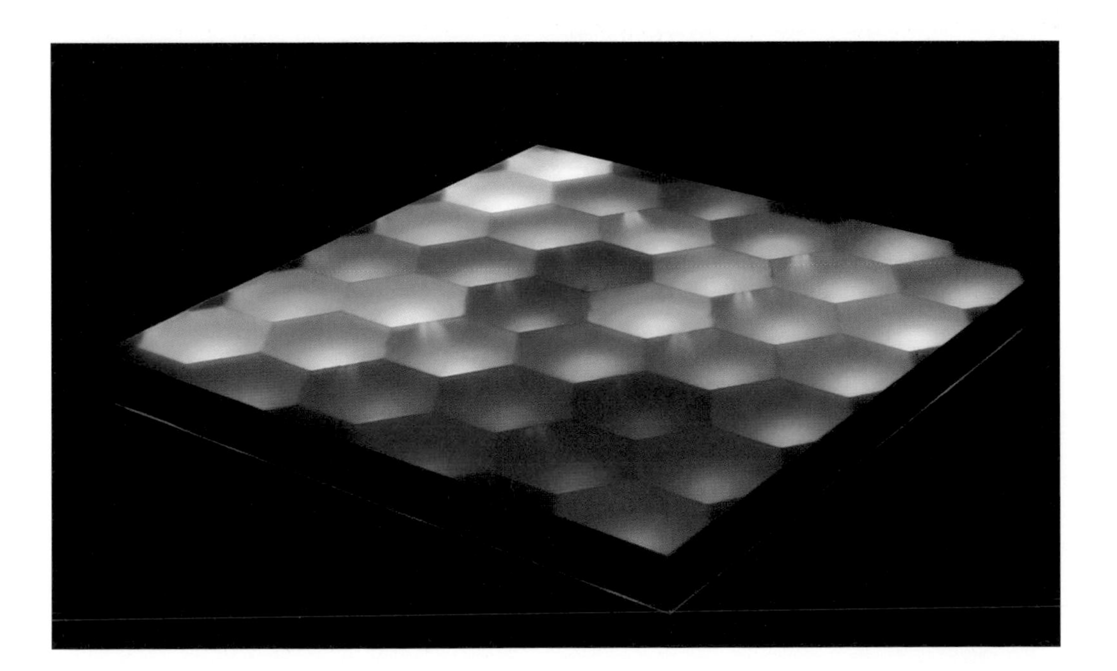

introducing
smart

The two merged into his thought, 'Wouldn't it be perfect to have a digital display that was a building material?'

The building material started to emerge after playing around in his studio with some Christmas lights, which resulted in a honeycomb aluminium sheet being scanned along with the lights.

Setting up a separate company, called SmartSlab, and raising private finance for it, Barker saw it as a challenge to see if he could make 'this simple idea' a reality.

He's willing now to cautiously describe SmartSlab as 'on the way to a success'.

He sees SmartSlab bringing communication benefits to all sorts of environments from communal areas, cafés and clubs, shopping malls and also as changeable building fascias.

Barker, now head of Industrial Design Engineering at London's Royal College of Art, points out that SmartSlab isn't only a one-way display. 'You can send pictures and text from a mobile phone to the screen.'

**Honeycomb efficiency** The geometry of the honeycombs that bees build is highly efficient.

The comb is made up of hexagonal wax cells. The six sides of each cell are always aligned in a way that gives the cell two vertical walls, plus a 'floor' and a 'ceiling' both made up of two angled sections.

When the hexagonal cells come together to form the comb, it creates a structure that uses the least material for a given volume.

Similarly, the closed ends of the honeycomb cells are geometrically efficient. These are made up of three planes, pyramidal in shape. The dihedral angles of all adjacent surfaces measure 120°, the angle that minimizes the surface area for a given volume.

Nature does have some brilliant ideas.

# This TV show really smells. Great!



Our sense of smell is a powerful trigger to memory and emotion, and this was the motivation for Joel Bellenson and Dexster Smith to come up with DigiScents.

Their idea was to turn odours into digital data, and through this 'scentography' they hoped to transform entertainment, e-commerce, advertising, and communication in general, into a fuller, more emotionally charged experience.

Based on psychological tests, it's thought that human beings can detect between 4,000 and 10,000 different smells. Each one is made up of various concentrations of the seven primary odours. These are ethereal (e.g. dry-cleaning fluid),

camphoraceous (e.g. mothballs), musky, floral, pepperminty, pungent (e.g. vinegar) and putrid (e.g. rotten eggs).

DigiScents involved a computer linked up to a small black plastic box. The box contained small vials of oil and a fan. The computer's signal heated the appropriate oil and the fan wafted the aroma out of the box into the immediate environment – and into the nostrils of anyone present.

It was easy to imagine a really wide range of applications for DigiScents, with everything from feature films to school history lessons being made a richer experience through olfactory enhancement.

For the moment, however, all this is on hold, as the DigiScents company is no longer operating. But the thinking remains, of course.



**Why so emotive?** Odour signals travel from our nostrils to an extension of the brain's limbic system along unique pathways.

The limbic system processes our emotions and spontaneous reactions, so a smell can cause a quick, involuntary response, such as salivating at the smell of freshly baked bread, or gagging at the smell of rotting flesh.

The limbic system also plays a part in the storage and recall of memories, and in the processing of sexual urges. No wonder both are so influenced by smells. And no wonder people selling their homes are sometimes advised to bake bread when a prospective purchaser is about to visit.

# Lose your phone without losing all your friends.

Intelligent Mobile Phone Charger
FINALIST, 2002

It's annoying when you lose your mobile phone, but it's replaceable. Often, the bigger loss is all the contact numbers you've stored on your phone, because some of these are probably irretrievable.

Paul Stokes' invention was sparked when a friend of his, Derek McGovern, suffered this experience.

Stokes' Intelligent Mobile Phone Charger is perhaps blindingly obvious – but he thought of it first.

His idea was to connect the phone's memory, via a small wire concealed in the charger cable, to a memory chip placed in the charger. Every time the phone was connected to the

charger, its memory would be automatically integrated, and any new numbers would be downloaded and added to the existing data.

The Intelligent Mobile Phone Charger has yet to see the light of day. The indifference of developers is not an unusual experience for inventors, but in this case it doesn't seem very intelligent.



**Some phone facts** Few inventions match the mobile or cell phone for blessing and blight. In an emergency it could be a lifesaver, potentially close to the opposite when it rings in a theatre.

The British scientist Michael Faraday probably started it all when, in 1843, he began researching whether space could conduct electricity.

In 1865, a dentist, Dr Mahlon Loomis of Virginia, may have been the first person to send a message through the atmosphere. He got a $50,000 research grant from Congress to look further into wireless communication.

Dr Martin Cooper has yet to reach the legendary status of Alexander Graham Bell, but he is considered the inventor of the portable handset, in 1973. He was also the first person to make a call on a portable cellular phone.

Four years later, cell phones went public in the USA.

Now, less than 30 years later, it's hard to imagine how we ever coped without them.

# Watched any good building blocks lately?

Digital Building Blocks
FINALIST, 2000

In 1999, New York's Museum of Modern Art exhibited 'The Digital House', designed by the Iranian-born architects Gisue Hariri and her sister Mojgan Hariri.

The Digital House showed off the attributes of the Digital Building Block.

The Block is a thin, transparent display, capable of transmitting and receiving data. When it's connected to the Internet, it enables communication, education, entertainment and work to take place within one single wall.

When it isn't 'switched on', the Digital Building Block will be like a sheet of clear glass. It can replace all computer

and television monitors, whether their functions relate to home or to work.

Gisue Hariri predicts that this thin, smart, transparent block will change our working and living spaces to more efficient, less cluttered and more flexible environments.

A Digital House would be organized around a touch-activated digital 'spine', which is, in effect, the brain of the house. It would work as a blank canvas, capable of displaying images, advertising, text, colour or any other digitized information. Alternatively, it could serve simply as a glass enclosure allowing contact with the landscape outside the house.

The various living spaces of the house would be plugged into the digital spine. The individual functions of the dining, living, sleeping and working areas of the house would benefit from the very different experiences the spine could offer.

Imagine preparing a meal in the kitchen with the help of the chef from your favourite restaurant. Or in the dining room, having dinner with a virtual guest.

In the living areas the spine could be an amazing entertainment centre.

Digital Building Blocks is a concept that is not yet a reality, but there's already an impressive virtual order book for it.

**Glass** Glass is remarkable, but its main ingredient couldn't be more unremarkable. It's sand. Add some soda ash and limestone to it, heat it to about 1,500°C (2,730°F), let the molten, red-hot mass cool, and you've made glass.

Different types of glass are made by adding other ingredients. Lead oxide makes lead, or crystal glass. Add boron and you get heat-resistant glass for cookware, and so on.

The colours in stained glass come from adding compounds of metals like copper, nickel, chromium and cobalt.

Sheet glass is made by floating a thin layer of molten glass over a bath of molten tin. The surface of the tin is perfectly flat, so the glass layer is, too.

# See the light. Charge a battery.

Solar-Powered Mobile Phone Charger
FINALIST, 2002

Invention is frequently about making indirect mental connections.

Richard O'Connor's Solar-Powered Mobile Phone Charger was inspired by his concept for a solar-powered smoke alarm.

The charger is a small, self-contained unit – a battery in fact – that allows anyone, anywhere, at any time to charge up their mobile phone, PDA, or even Walkman or MP3 player.

What distinguishes O'Connor's invention is this: it gets its energy from both light indoors and outside. Its revolutionary charge-control circuit and new technology allow charging at lower light-levels than ever before.

Being left in bright direct summer sunlight for just two days will fully charge the unit. And once fully charged, it will keep its charge for a minimum of five years – and that's without solar energy.

The benefits are wide-ranging. There's the day-to-day assistance to mobile-phone users, from adults in the developing world to schoolchildren in the developed world, on the move and without access to mains or car charge. There's also the fact that the charger is financially and environmentally economical. Running costs are low, and very little battery waste is produced compared to conventional batteries.

A charger with all the positives and no negatives?



**The Sun** The Sun, 148.72 million km (92.95 million miles) from Earth, is our nearest star. Its mass is 330,000 times that of the Earth, with a diameter of around 1.4 million km (875,000 miles).

The Sun is about 70 per cent hydrogen and about 28 per cent helium. (The remainder is oxygen and carbon.)

The Sun's energy is generated by nuclear fusion reactions in its core, and it converts 4.3 million tonnes of itself into energy every second.

The temperature on the surface of the Sun is 5,500°C (9,900°F). Quite cool compared to its core temperature of 15,600,000°C (28,000,000°F).

The Sun radiates as much energy in 15 minutes as the entire human race consumes in a year.

# Don't waste your energy walking.

Biomechanical Transducer/Electric Shoe
FINALIST, 2000

Trevor Baylis is probably the UK's best-known living inventor, and he leads a crusade on behalf of fellow inventors who, all too often, find themselves thrashing around in shark-infested commercial waters.

To help other inventors get their ideas to market, Baylis has set up a commercial organization, Baylis Brands.

Baylis, who invented the hugely successful clockwork radio, has turned his hand to feet.

Collaborating with John Monteith and Barry James, Baylis set out to recover the energy that's wasted when walking.

It used to be the case that most electrical gadgets needed

more energy than the human body can comfortably generate in a short length of time.

The three men investigated a number of options presented by the human body to identify an action that would create a significant amount of electricity efficiently. They also considered how this power, once generated, might be stored.

The action of walking emerged as the front-runner, and they invented the technology needed to convert into electricity the energy used to walk. The technology was designed to be housed in the sole of the walker's shoe, boot or overshoe.

Their Biomechanical Transducer emerged after exploring many other possibilities, using pneumatic, electro-thermal, hydraulic and other technologies.

In spite of the obvious potential benefits of the Electric Shoe to people without funds or access to electricity, it has yet to see the light of day.



**Screaming jelly babies** We get our own energy from the food we eat. Calculating energy in a particular food is done by completely burning it in a calorimeter; the heat released is accurately measured to determine the food's gross energy value.

A dramatic chemistry experiment is performed in schools across the UK to demonstrate how much energy there is in a jelly baby.

Potassium chlorate is melted in a test tube and a jelly baby is dropped in. The jelly baby bursts into flames and makes a very nice screaming sound as it burns, accompanied by the smell of candy floss.

# How to charge up a phone without plugging it in.

Splashpower
FINALIST, 2005

The race to develop more and more sophisticated mobile communication devices has left battery technology looking decidedly primitive.

Mobile devices with features such as video, MMS and interactive gaming require high-speed wireless networks, backlit colour screens and more powerful microprocessors. This has made their power consumption increase dramatically and has created a power gap between their needs and what present battery technology can deliver. The power gap is also constraining the development of next-generation devices.

Our reliance on mobile communication devices quickly

OPTICAL
4.0

shifts from frustration to anger and mild panic when we run out of power in a situation where it's impossible or inconvenient to recharge. Then there's the business of different, dedicated chargers for each device.

Splashpower cuts through all these problems and meets the needs.

Splashpower's unique technology enables a mobile phone and other portable devices to be charged up at the same time by placing them on a mouse-mat-sized pad, called a SplashPad.

The devices don't have to be plugged in because the power is transferred wirelessly from the SplashPad.

Splashpower technology enables the necessary power to be delivered to charge typical portable devices, placed in any orientation over the active charging surface of the SplashPad. They're charged up at the same rate as they would be when plugged into a conventional charger. Devices with different power requirements can be charged at the same time.



**Dead frog gives birth to first battery** Batteries work through the reaction when two pieces of metal (electrodes) make contact with a chemical (electrolyte).

In 1791, Luigi Galvani, an Italian anatomist, was examining a dead frog's nerves. The instruments he was using were made of different metals, and he noticed the frog's leg muscles twitching.

When Alessandro Volta heard about Galvani's discovery he worked out that electricity caused the twitching.

Galvani's instruments had played the part of a battery's electrodes, the frog's bodily fluids the part of its electrolyte.

So Volta was inspired to create his 'voltaic pile' in 1799.

Made up of a number of cells, each with a zinc and a silver electrode with a layer of card soaked in salt water between them, Volta's invention was the world's first battery.

# Robots can have feelings too, you know.

The whole story of Quantum Tunnelling Composites sounds closer to fiction than fact.

What started as the search for an electrically conductive adhesive for use with a new type of monitored security tag, somehow ended up at NASA on the fingers of their robots' 'hands'.

How did that happen?

David Lussey set up a security business after leaving the British Armed Forces.

In his home workshop in Richmond in England's north-east, he was trying to make the aforementioned adhesive, which would show if any illegal attempt was made to peel the tag from the item it was protecting.

He was mixing different adhesives with different conductive powders and testing them for their electrical qualities. One mixture showed an electrical result that defied logic. When he stuck two aluminium foil strips together with the mixture, the joint showed no electrical conductivity until he pulled the strips. Then it became very conductive, but only while he continued to pull.

Not being a scientist himself, he took the mystery mixture to Professor David Bloor, a physicist at Durham University, who confirmed that it was an entirely new form of conductive substance. It looks similar to the rubber in a car tyre.

Its unusual mode of conduction, which scientists call quantum tunnelling, led to the name QTC (Quantum Tunnelling Composite).

In its normal state, QTC is a virtually perfect electrical insulator, which can be transformed into a metal-like conductor by applying an extremely small force or pressure. Take the pressure off and QTC goes back to being an insulator.

The potential range of uses for QTC is vast. New types of electrical controls, switches, keyboards and pressure sensors are just the beginning.

QTC's peculiar properties are already being applied in keyboards, pressure-sensitive cables and gloves, and many

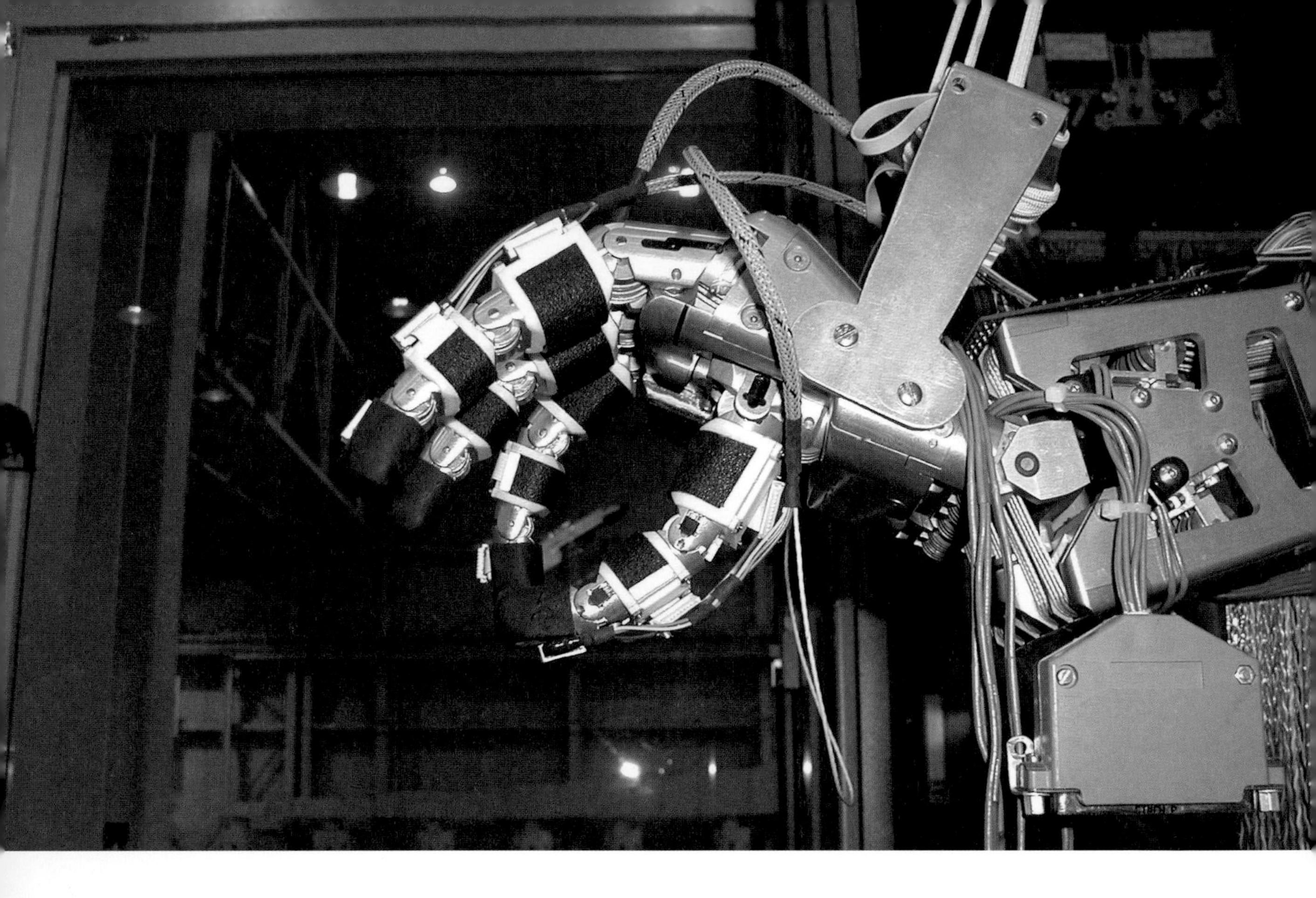

new products are in the pipeline – from clothes and toys to sports goods – around the world.

And QTC has applications out of this world, too. As a sensing technology it is unique. Because it can detect or emulate human touch, NASA is now using it on the fingers of robotic hands to give them a human-like sense of touch.

QTC's sensitivity is such that it can also detect airborne vapours and gases in very low concentrations.

Peratech, the company David Lussey set up to make and market QTC, is able to produce 10 kg (22 lb) a day, which is enough to make up to 100,000 switches or controls.

While 150 university students have carried out commercially orientated research on QTC, Peratech is exploring and encouraging further applications for their composite by selling it through a UK high-street retailer and in trade outlets.

**Quantum tunnelling** Standard conductive composites are usually made from polymers filled with carbon. Some of the carbon particles are always in contact with one another, creating a conduction path. When pressure is applied to this kind of composite, more carbon particles come into contact, so more conduction paths build up. This process is known as percolation.

By contrast, in QTCs, the metal particles never come into contact, but they do get very close. So close, in fact, that quantum tunnelling is possible between the metal particles.

Quantum tunnelling derives from quantum mechanics, in which an electron isn't viewed as a solid particle but more like a wave. When the 'wave' meets a barrier, a non-conductive material, for example, it doesn't instantly go to zero, but decays exponentially.

If the 'wave' hasn't reached zero by the time it has gone through the barrier, then it emerges on the other side. In other words, the electron has effectively 'tunnelled' through the non-conductive barrier.

# INTO CYBERSPACE

'Cyberspace' the word, was invented by one of our first award judges, the author William Gibson.

'Out of Control' was the title of Kevin Kelly's most recent book when we met for lunch at a small restaurant in San Francisco. We were particularly interested in Kevin Kelly's point of view on the launch of the award.

Apart from being an innovative thinker, Kevin was also a champion of new ideas as co-founder of Wired magazine.

Could we get Wired to do a supplement on the finalists?

It felt like absolutely the right medium for this.

Even the layout of the magazine was innovative. As Kevin described it to me the original design brief they'd set themselves was "You receive in your mailbox a magazine from the future. What would it look like?"

As it turned out the Wired lead time wouldn't work for us. That meant a re-think on what had been our hopeful launch strategy.

We decided on another wired approach. A global live braadcast.

The Saatchi & Saatchi Innovation in Communication Award was launched on 7th October 1997. The webcast was hosted simultaneously by 83 Saatchi & Saatchi offices in 132 countries.

Present were press and TV journalists from each country and a panel including Buzz Aldrin, James Burke and Trevor Baylis in London, and Edward de Bono in Sydney.

Each panelist talked about the significance of the award.

Edward de Bono said "We are leaving the information age and entering the age of creativity. Competence, information, and technology are becoming commodities. What will differentiate companies in the future will be value creation, and that demands new ideas."

Kevin Roberts, my partner and Worldwide CEO, announced to Edward's point that "From today we are taking 'Advertising' out of our name around the world and re-structuring as an 'Ideas company!'"

This caused some confusion and derision within the ad world. As it turned out though, we were 7 years ahead of a curve brought about by an explosion of new media outlets, not the least of which was the internet.

The Innovation in Communication Award was probably the first time the web had been used as the primary media to drive a mainstream press launch.

5

# The Medical World

# How to stop medicine killing patients.

Universal Pill Marking System
FINALIST, 2000

Prescription drugs, designed to cure, can kill. Because the right person has to take the right dosage of the right drug at the right time, mistakes are a significant cause of death and illness.

Richard Peterson and the late Dr Gerhard Ovellhaus found a way of reducing the mistakes. They developed a system for marking pills, capsules and other solid-form pharmaceuticals with an identifying mark that combines an intuitive, non-linguistic symbol with a machine-readable barcode.

So the information would be recognizable both by patients, regardless of language or education, and by healthcare workers,

Heart

Heart & Blood

Diamond

Wellbeing

Spade

Sex Organs

Club

Stomach

Cross

Emergency

Triangle

Nutrition

Square

Blocker

Star

Pain

Sun

Head

Tree

Muscles & Bones

Fish

Eye

Crown

Brain

Hand

Extremities

Castle

Skin

Target

Psyche

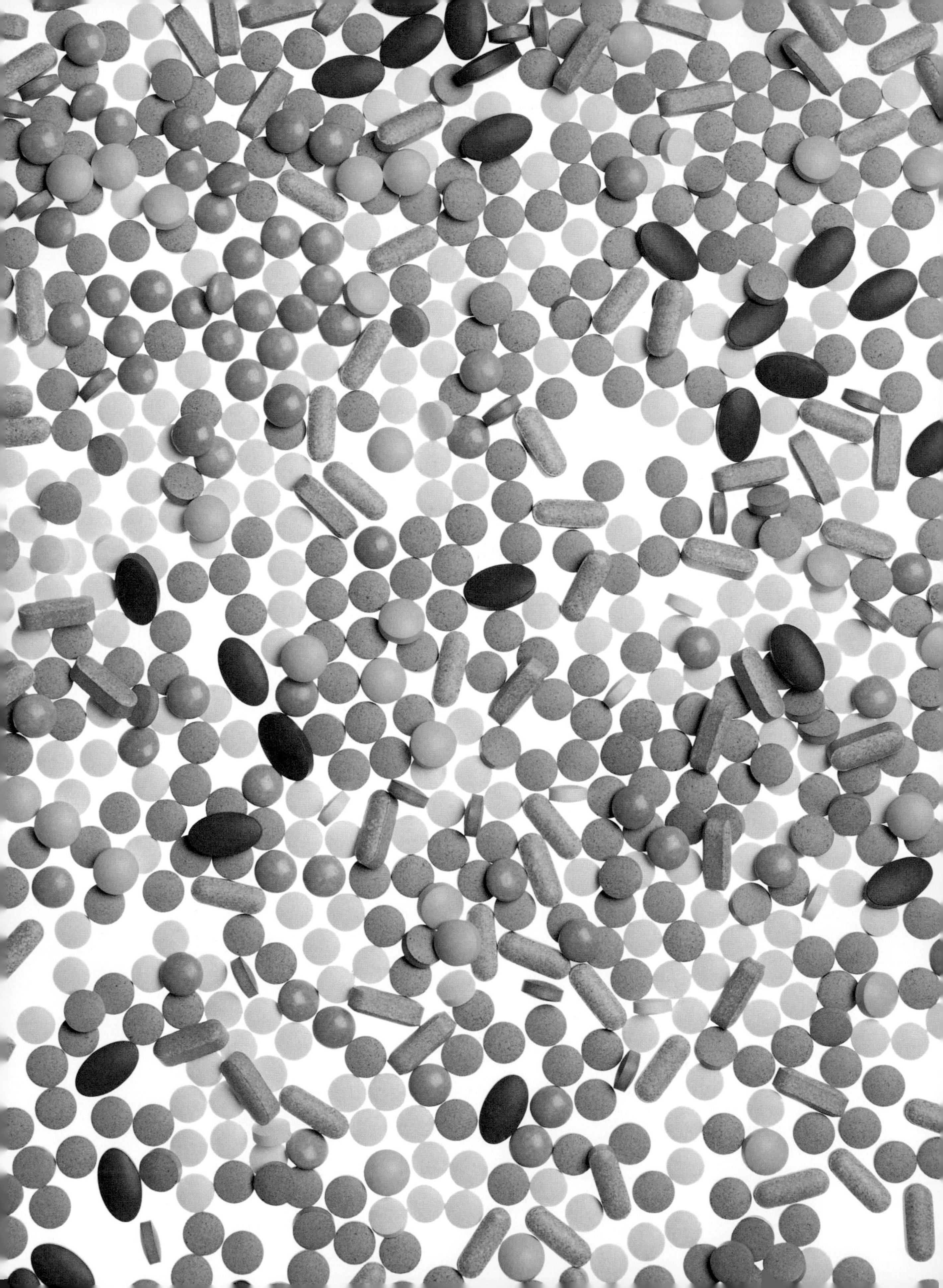

the symbols which include, for example, a heart, cross, and star, are intended to suggest the condition being treated. Although small, the symbol would be recognizable and would appear alongside one or more numbers or letters identifying the drug in terms of substance, dosage and manufacturer.

The pill would also be marked with a barcode, readable by a simple portable code-reader.

The idea, which has yet to be adopted, is now in the hands of a Pacifica, California company, Quintal.

The company describes itself as 'a loose-knit group of oddballs and blockheads but serious about licensing technology that should make people's lives easier'.

Or in this case, saves people's lives.



**Symbolic power** Powerful non-linguistic symbols can speak volumes. They can go way beyond communicating basic information, to capturing whole associative histories and triggering the strongest emotions.

Their significance can also be totally transformed by misappropriation and re-application.

Perhaps the most awful example of this is the swastika. In its ancient and original form, the swastika represented cosmic dynamism and creative energy. Its name comes from the Sanskrit words 'su', meaning 'well', and 'asti', meaning 'being'.

The symbol's essential meaning – of life force, solar power and cyclic regeneration – is often extended to signify the Supreme Being, as in Jainism.

The swastika's shape suggests a certain whirling momentum. But its shape also divides space into quarters, so it has associations with the four wind gods, the four seasons and the four cardinal points.

Unfortunately, the swastika now has associations only with Nazi Germany.

# Intensive care where it's really needed. Nowhere near a hospital.

LSTAT (Life Support for Trauma & Transport)
FINALIST, 2000

The first 60 minutes after a serious injury is called the Golden Hour, and what happens to the patient during this time can mean the difference between life or death.

The LSTAT, the Life Support for Trauma and Transport patient-care platform is, in effect, a portable intensive-care unit. It's a 13-cm (5-in) thick 'smart' stretcher, but no larger than a conventional one, weighing only 75 kg (165 lb).

The LSTAT incorporates a ventilator, with on-board oxygen machine, physiological monitor, defibrillator, surgical suction unit, a three-channel fluid and drug-infusion pump, and a blood-chemistry analysis function. On-board batteries supply the field power and all functions are connected to a data-logging and communications system.

The LSTAT is able to provide an unbroken continuum of care from where the injury is sustained, through transport, surgery and recovery. If a patient's breathing becomes erratic, an alarm sounds on the LSTAT to alert the medical team. The medics can monitor blood pressure and oxygen levels and set up a drip, leaving the LSTAT to administer the drugs. The LSTAT can administer three types of drugs: pain relief, cardiac support and anaesthesia. If the patient's heart stops, the defibrillator can restart it, and, if necessary, tell an inexperienced user how to do it.

Although the LSTAT was developed initially for US military use, it has been adopted by the military medical communities of several nations, and evaluations with civilian hospitals and ambulance companies have already begun.

According to the American Trauma Society, traumatic injury is the primary cause of death in the first half of life anywhere around the world, and the World Health Organization sees it as a global epidemic.

Integrated Medical Systems Inc., the makers of LSTAT, picked up the United Nations-sponsored Tech Award for 'technology benefiting humanity'.

Dr Matthew Hanson of IMS sums up LSTAT by saying, 'For the first time in history, patient data from a number of

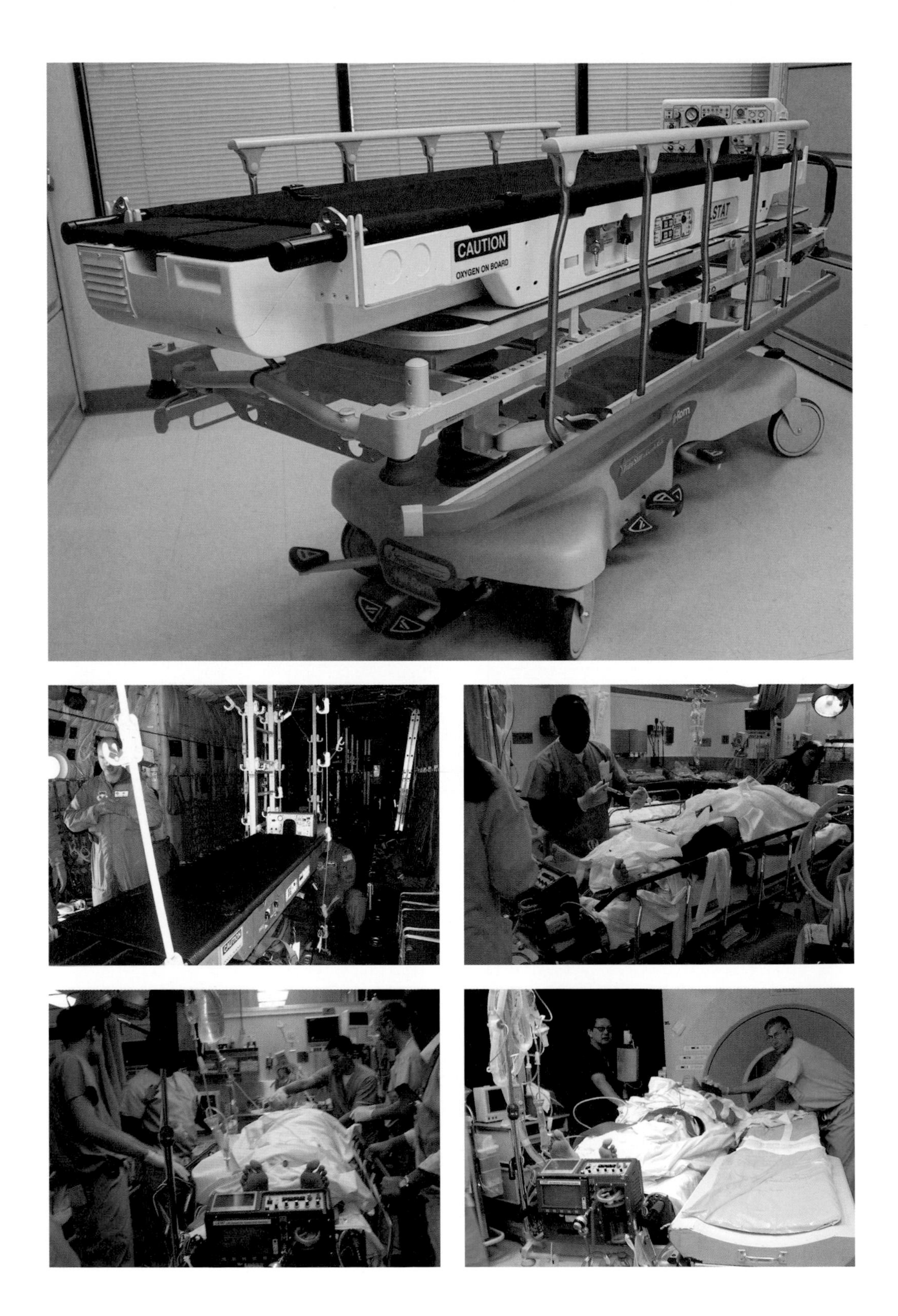
CAUTION
OXYGEN ON BOARD
LSTAT

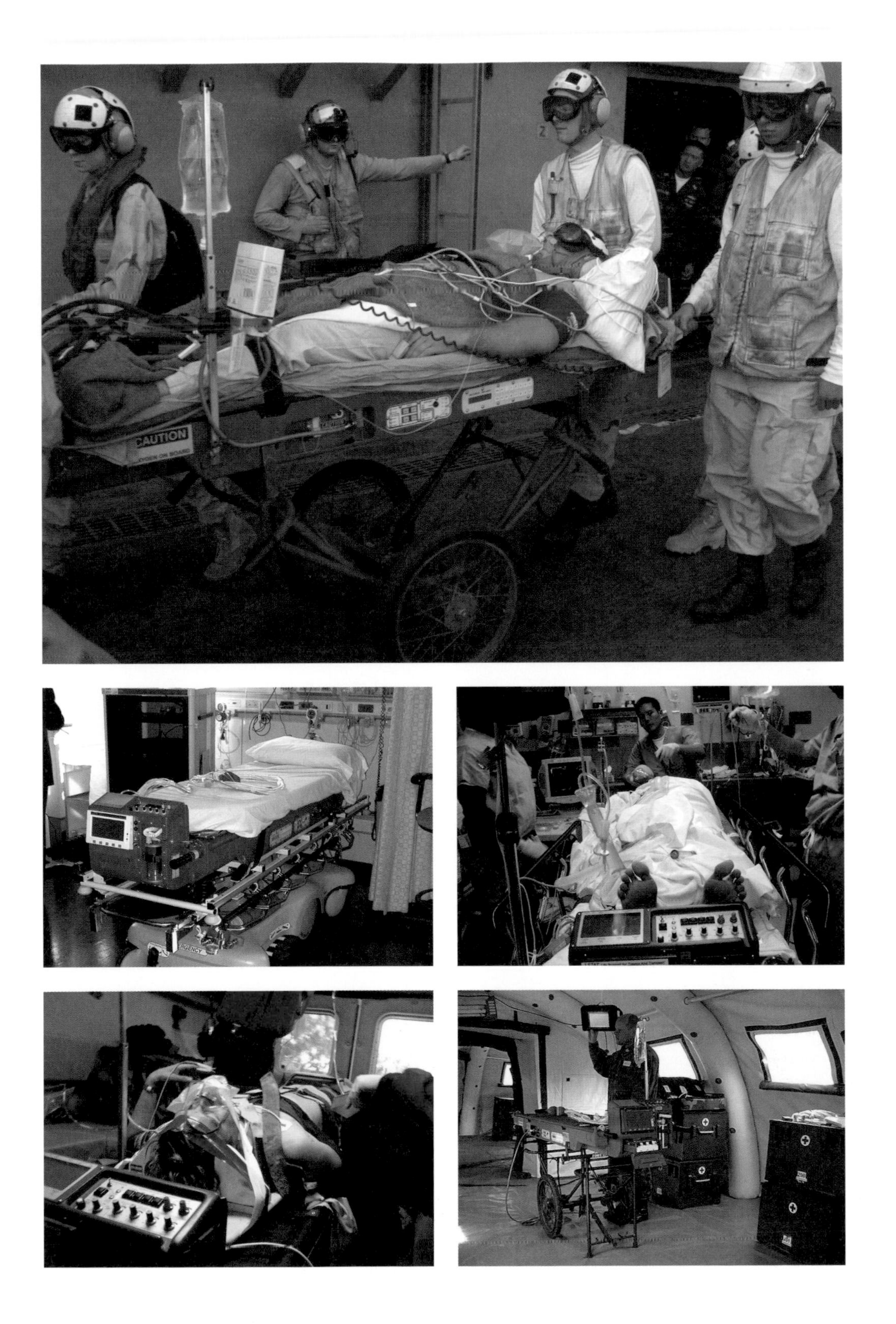


LSTAT (Life Support for Trauma & Transport)
FINALIST, 2000

medical devices is time-synchronized and made available in real time to anywhere in the world. The LSTAT platform allows a conscious or unconscious patient to communicate across space and time with their own body, with medical equipment and with caregivers and researchers near and far.'

You can find LSTAT at www.lstat.com.



**The Golden Hour** The theory of the Golden Hour was the brainchild of R. Adams Cowley, who founded the Maryland Shock Trauma Center in Baltimore. He identified this hour of opportunity in which the lives of severely injured people could be saved with the appropriate treatment.

Two-thirds of those who die as a result of serious trauma will have suffered major head or other central nervous system injuries and little could have been done to prevent their deaths. However, 66 per cent of the remaining fatalities would be preventable if the casualty was to receive appropriate medical management in the Golden Hour.

The majority of preventable deaths in these circumstances are the consequence of inadequate airway management. The rest can be put down to inadequate management of ineffective breathing or poorly treated shock following haemorrhage.

Modern emergency medicine owes a lot to discoveries and techniques developed in field hospitals in the Korean and Vietnam wars. Doctors in these MASH units were saving the lives of people who would have died in World War II. The skills and knowledge acquired in war found a very valuable application in peacetime.

# The machine that reads minds.

NeuroGraph Diagnostic Aid System
FINALIST, 2002

The NeuroGraph Diagnostic Aid System can detect the early signs of Alzheimer's disease, before there are any behavioural symptoms.

It can also help patients suffering from depression by quickly determining the most effective anti-depressant for them, rather than going through a protracted period of trial and error with a variety of different drugs.

The NeuroGraph is a portable, non-invasive device that reads complex activity in the brain. It uses an electrode-wired cap, goggles and headphones for delivering stimulus material, and a low-noise amplifier.

DB Generator
B) Load DB File
C) analyzer
Transform script
(6) Report Generator
16/23
20/23
18/23

The initial screening takes only about 20 minutes, and the speed and accuracy of the NeuroGraph (nearly 100 per cent) in identifying neurological disorders is invaluable, especially for those showing early signs of Alzheimer's.

There's substantial evidence to demonstrate that earlier treatment of many brain diseases can postpone the most serious symptoms.

This is particularly important in cases of Alzheimer's, as previously there was no sure-fire way to identify the disease short of a brain biopsy – a procedure usually undertaken only after death to confirm the suspicion of it.

Before the NeuroGraph, diagnosis of Alzheimer's, schizophrenia and similar diseases relied on questionnaires, reaction times and psychiatric observation. Or expensive, cumbersome and slow imaging systems.

The NeuroGraph is relatively inexpensive, at around £190 ($350) per diagnosis, and eliminates the interpretative guesswork of non-technological methods of diagnosis.

The NeuroGraph will also play a role in the clinical trials of new drugs by monitoring their effect on the patient's brain activity.

The NeuroGraph, like many inventions, did not originate directly from the inventor looking for a way to diagnose diseases like Alzheimer's.

Dr Richard Granger was the director of the University of California's Brain Engineering Laboratory. His studies into how our brains can recognize sights and sounds led him to build a brain-derived computer system for the US Navy that processed naval sonar signals.

Then he realized that he could apply these methods to other sorts of signals and could single out key brain signals that correlated with particular disease systems.

Dr Granger is a director of the Thuris Corporation, the makers of the NeuroGraph. He is also Distinguished Professor of Computational Sciences at Dartmouth College, and Professor of Psychological and Brain Sciences and the director of Dartmouth's Neukom Institute for Computational Sciences.

**Alzheimer's disease** This serious form of dementia gets its name from the Bavarian neuropsychiatrist Alois Alzheimer, after he described the symptoms at a meeting of psychiatrists in Tübingen, Germany on 4 November 1906.

He described the seemingly insane behaviour of one of his patients, Auguste D. – her worsening memory, her disorientation, her non-sequiturs, her occasional delirium and hallucinations, and her frenzied jealousy towards her husband.

When Auguste D. died, a post-mortem examination of her brain revealed vivid peculiarities to Alzheimer. Thousands and thousands of tiny clusters were strewn across her cerebral cortex. Alzheimer also noticed innumerable nerve cells whose interiors were choked by 'dense bundles of fibrils'. These knots are known today as 'neurofibrillary tangles' and are associated with several other brain diseases.

# To beat cancer you have to think small.

Optical Stretcher
FINALIST, 2005

Cells are microscopic building blocks, and each of us has an estimated 10 million million of them making up our bodies.

Cells are informative too, and by observing changes in their make-up they can be used diagnostically.

Until now, the observation of cells has focused on their biological make-up.

The Optical Stretcher is different. It uses the physical characteristic of cells – their stretchiness or elasticity – to observe small changes.

The Optical Stretcher does this by using the forces created by powerful beams of infrared laser light to stretch and

When a cell is positioned between two opposed, non-focused laser beams, it is stretched along the axis of the beam. Cancer cells can be identified because they are more elastic than healthy cells.

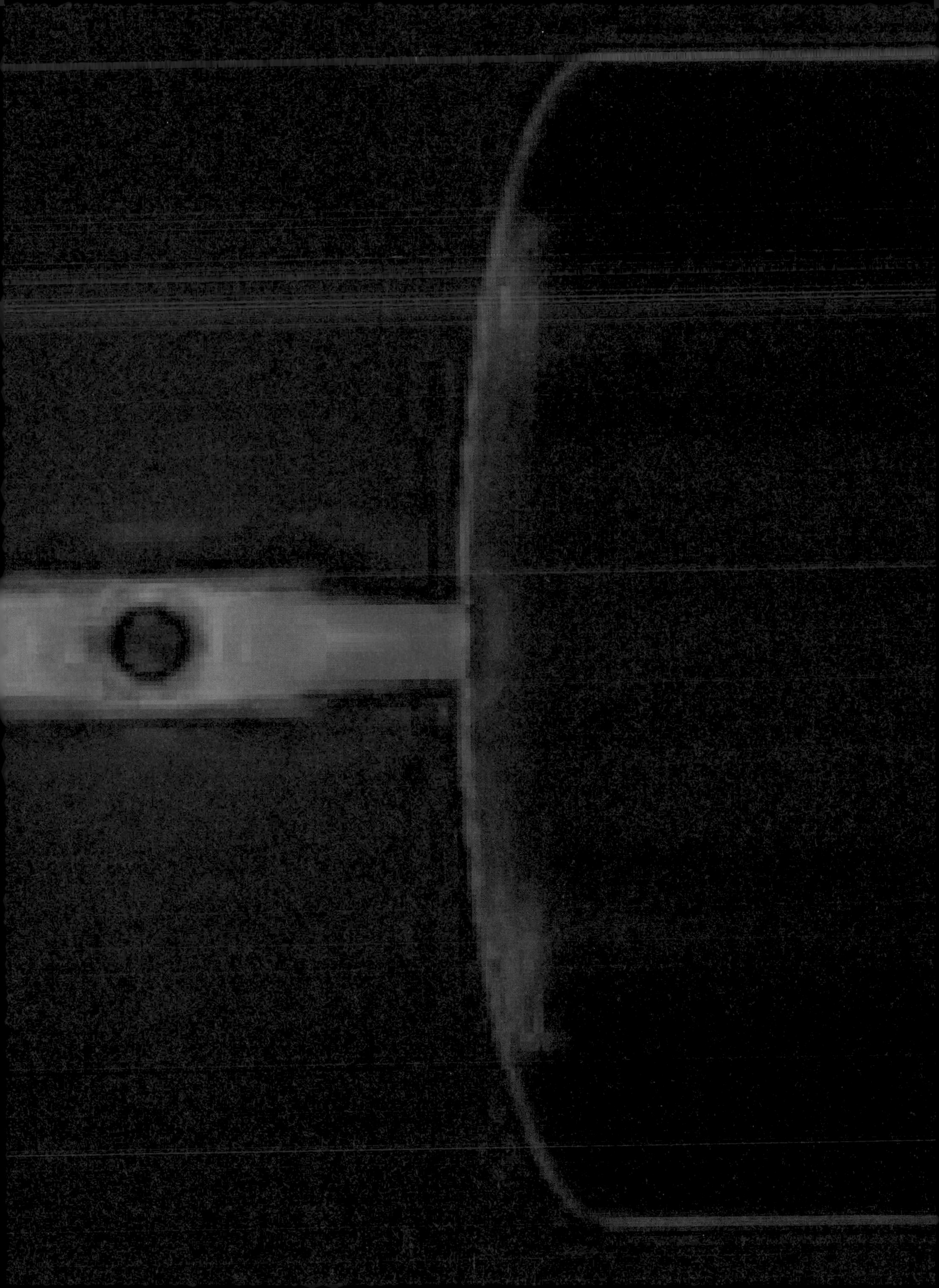

measure cells one by one. It does this without killing the cells.

This process is highly effective for diagnosing cancer and how it may be spreading. Of all the physical properties of a cell, elasticity is the one that varies most dramatically between normal and cancerous cells. This makes stretching the most sensitive method for identifying cancer.

And just 50 tumour cells within a sample are required, compared with the 10,000 to 100,000 needed for traditional testing. So instead of invasive biopsies, doctors can extract a sufficient sample using only a fine needle.

The Optical Stretcher can also determine, just by measuring cells from the primary tumour, whether or not the cancer is likely to be spreading.

Secondary tumours can be difficult to find, and women with breast cancer often undergo precautionary mastectomy or whole-body chemotherapy.

The emotional and physical impact of these procedures could be avoided when the Optical Stretcher is used.

Where cancer cells lose their rigid cytoskeleton, making them more elastic, stem cells are more elastic because they don't have cytoskeletons to begin with. Stem cells are cells that have not become the building blocks for specific use in the body, such as for skin, brain or muscle, and they could hold the key to clues for currently incurable conditions, such as Parkinson's, heart disease or spinal-cord injuries.

The promise of finding these cures is offset by the controversy surrounding the use of embryonic stem cells. It is possible to isolate stem cells from adults, but current methodology is expensive and often contaminates the cells.

But because stem cells are stretchier than normal cells, the Optical Stretcher can identify and sort them, and it's possible that, in future, it will play a pivotal role in stem-cell-based regenerative medicine.

The inventors of the Optical Stretcher are Doctors Josef Käs and Jochen Guck.

**Lasers** 'Laser' stands for 'Light Amplification by Stimulated Emission of Radiation'.

So, in a laser, a light source is made stronger by rays given off by atoms that have been given extra energy.

The first laser, the ruby laser, was invented in 1960. A bright flash produced by a high-intensity light started the laser. Atoms of chromium inside a rod of artificial ruby, excited by this light flash, emitted red light.

Laser light differs from the light emitted by a bulb in several ways. Laser light travels in the same direction (coherent light). Bulb light travels in all directions.

Light from a bulb is a mixture of light of many different colours. Laser light is a single, pure colour. Light from a bulb spreads out too, so the further you are from it, the dimmer it seems.

Not so with laser light. This hardly spreads at all. In fact, there's a mirror the size of a tea tray on the Moon, left there by astronauts. This is used to reflect back a laser beam sent from Earth nearly 400,000 km (250,000 miles) away.

# For the inside story, swallow a camera.

Wireless Capsule Endoscope (Pillcam)

FINALIST, 2000

If the human body had straight intestines they would be easier to examine, but people would have to be nearly 10 m (33 ft) tall.

Because they're coiled up inside the abdomen, it's almost impossible for conventional diagnostic devices to examine the entire gut.

Gastroscopy can check the first 1.2 m (4 ft) of the upper digestive tract, and colonoscopy can evaluate the colon and rectum. This leaves about 6 m (20 ft) of small intestine beyond the range of regular devices and not entirely within the range of difficult-to-operate 'push enteroscopes'.

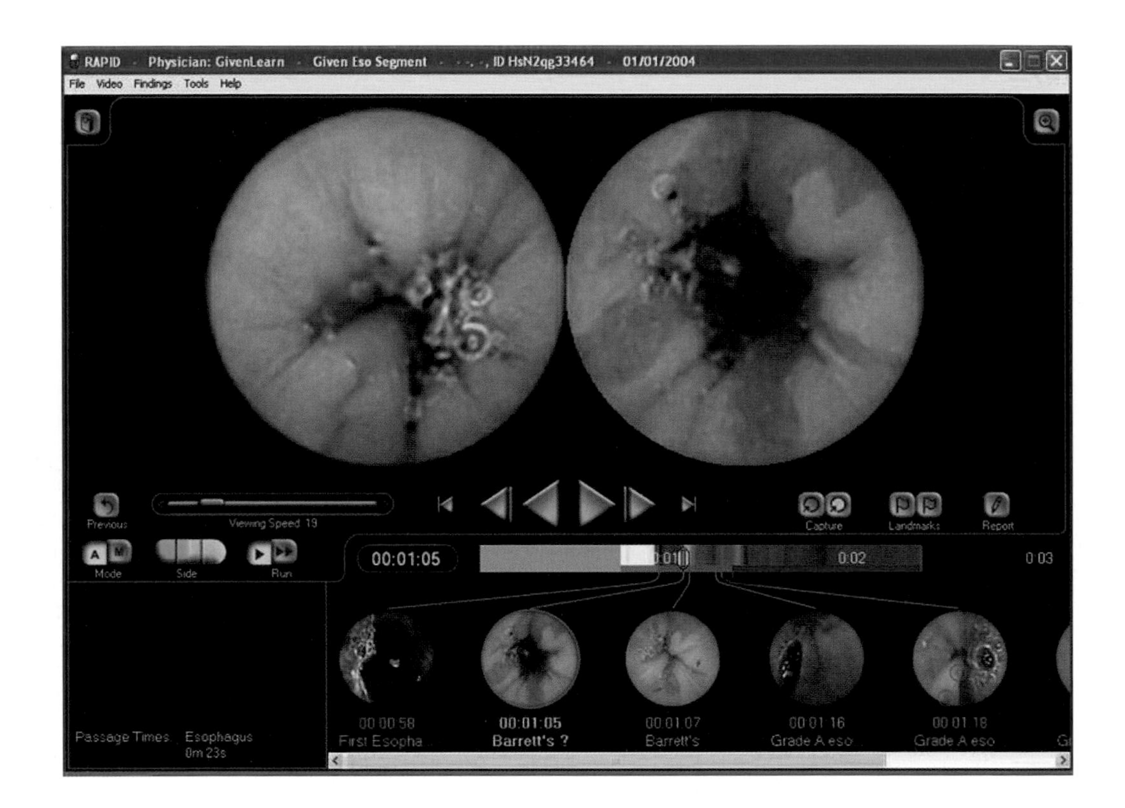


These images showing the condition Barratt's Oesophagus, were taken by a Pillcam model with two cameras. One looks forwards the other backwards.

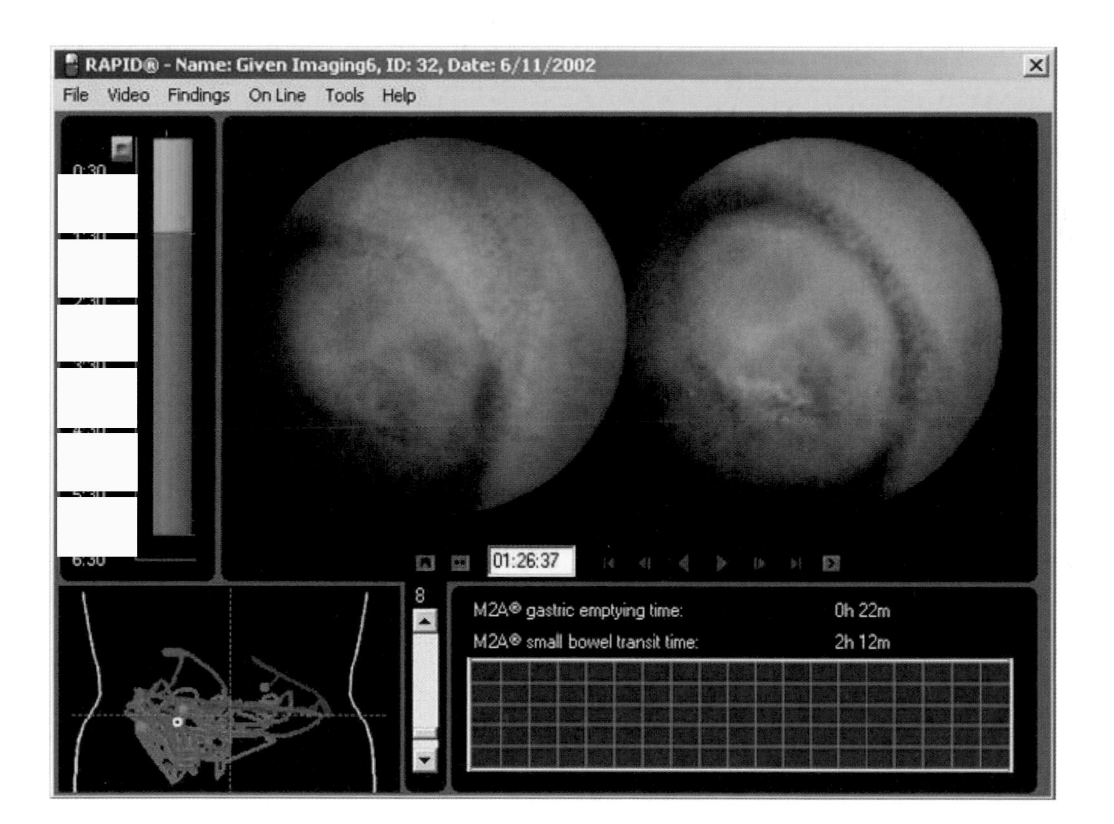


Pillcam images of a vascular tumour in a patient's small intestine. The condition is called blue rubber bleb nevus syndrome.

Which is where Wireless Capsule Endoscopy comes in. Jointly developed by a team in the United Kingdom and a team in Israel, the Wireless Capsule Endoscope is, in essence, a video camera in a pill.

Both teams had started working on the concept, separately, in the early 1980s, but it took 20 years for technology to catch up with their thinking.

In 1997, the two teams joined forces. Gastroenterologist, Professor Paul Swain had been working with physicist and bio-engineer Tim Mills and a team in London. Gavriel Meron and Gavriel Iddan had been working together in Israel.

In the true pioneering spirit of invention, Swain decided to be the guinea pig for the capsule, and swallowed the first one in a clinic outside Tel Aviv.

What exactly did he swallow?

A colour video camera and wireless radio frequency transmitter, four LED lights and enough battery power to take 50,000 colour images during its eight-hour journey through his digestive tract.

The 11 x 26-mm (0.4 x 1-inch) capsule, which is about the size of a large vitamin pill (shown actual size on page 159), weighs just 4 g (0.1 oz) and is made of specially sealed biocompatible material that is resistant to stomach acid and powerful digestive enzymes.

There's no need for sedation or hospitalization.

Patients say the camera is easier to swallow than an aspirin. It moves through their digestive tract naturally, helped on its way by the normal activity of the intestinal muscles.

The video camera's images are continuously transmitted to special antenna pads placed on the patient's body and captured on a Walkman-size recorder worn on the waist.

At the end of the examination, these recorded images are transferred to a PC workstation, where they are then converted into a digital movie for the doctor to examine.

The camera is disposable and is expelled naturally.

The invention is made by Given Imaging, and is called the Pillcam Capsule Endoscope.

**Endoscopic surgery** Two hundred years ago, if you needed surgery you'd also need to be brave and lucky. Lack of anaesthetics and antiseptics made the simplest operations agonizingly painful and extremely hazardous.

Fast-forward to endoscopic, or keyhole, surgery.

An endoscope tube is inserted into the patient through a small incision or a natural body opening such as the patient's mouth.

The tube houses bundles of optical fibres to illuminate the inside of the body and transmit images back to the surgeon.

Channels in the tube provide air, water and suction, and carry a range of tiny surgical instruments to the operation site, such as scissors, forceps, a cutting laser and a tiny brush to collect cell samples from tissue surfaces.

Guided by the image being transmitted from inside the patient, the surgeon uses the endoscope handset to manoeuvre the endoscope and to perform the procedure remotely.

Endoscopic surgery is quicker, reduces healing time and produces less scarring than open surgery.

## "THE ROLE OF BUSINESS IS TO MAKE THE WORLD A BETTER PLACE"

Kevin Roberts

Dr Irvine-Halliday is a Canadian of Scottish descent.

He wore a kilt on the evening we announced his Light up the World Foundation as the grand prize winner.

In the middle of his acceptance speech, this rugged man started to cry.

Several hundred people in the audience were deeply moved too as he explained the difference the prize money and our involvement would make to his work in the developing world.

As it turned out, it would exceed even his imagination at the time.

We continued to work with the Light up the World Foundation over the following year on a number of projects.

Then on 26th December 2004, a tsunami swept over Indonesia and outlying islands in one of the world's largest natural disasters.

Entire villages and townships were swept away and the number of homeless was staggering.

Our parent company, the Publicis Groupe, had immediately made a large financial donation, half to the Red Cross, half to be allocated by several of our managers in the region according to where they saw most need.

Temporary villages had been set up. They needed lighting.

Light up the world had been asked if they could help.

Dr Irvine-Halliday asked us the same question.

From the funds sent to our regional managers to distribute, we were able to help Light up the World provide their light emitting diodes to what Dr Irvine-Halliday later described as "probably and unfortunately long term temporary villages."

He sent us a letter of thanks so touching, it was his turn to bring a tear to the eye.

BOB ISHERWOOD

6

# The Disability World

# How the blind can see pictures with their fingers.



Photo-Form Tactile Graphics allow blind users access to a vast range of images that they 'see' with their hands and fingers.

They're bas-relief tiles created from any type of two-dimensional image using digital technology. The result, in hard, durable plastic resin, is a surface that accurately represents the original image with an emphasis on textures and forms.

This new method is radically different from traditional production, where string and other flat objects are placed on a board and a vacuum-forming process is used to produce a tactile image.

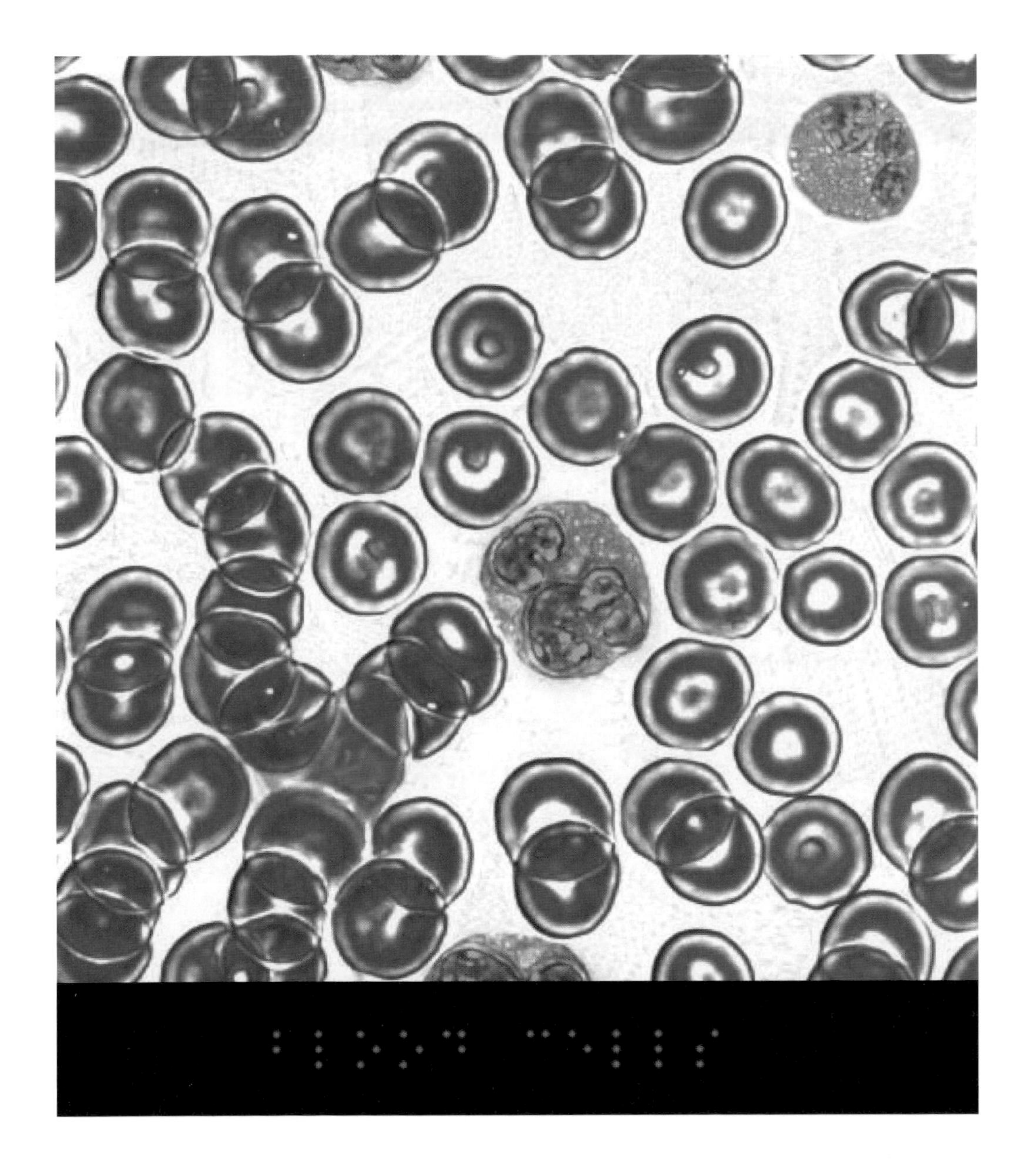

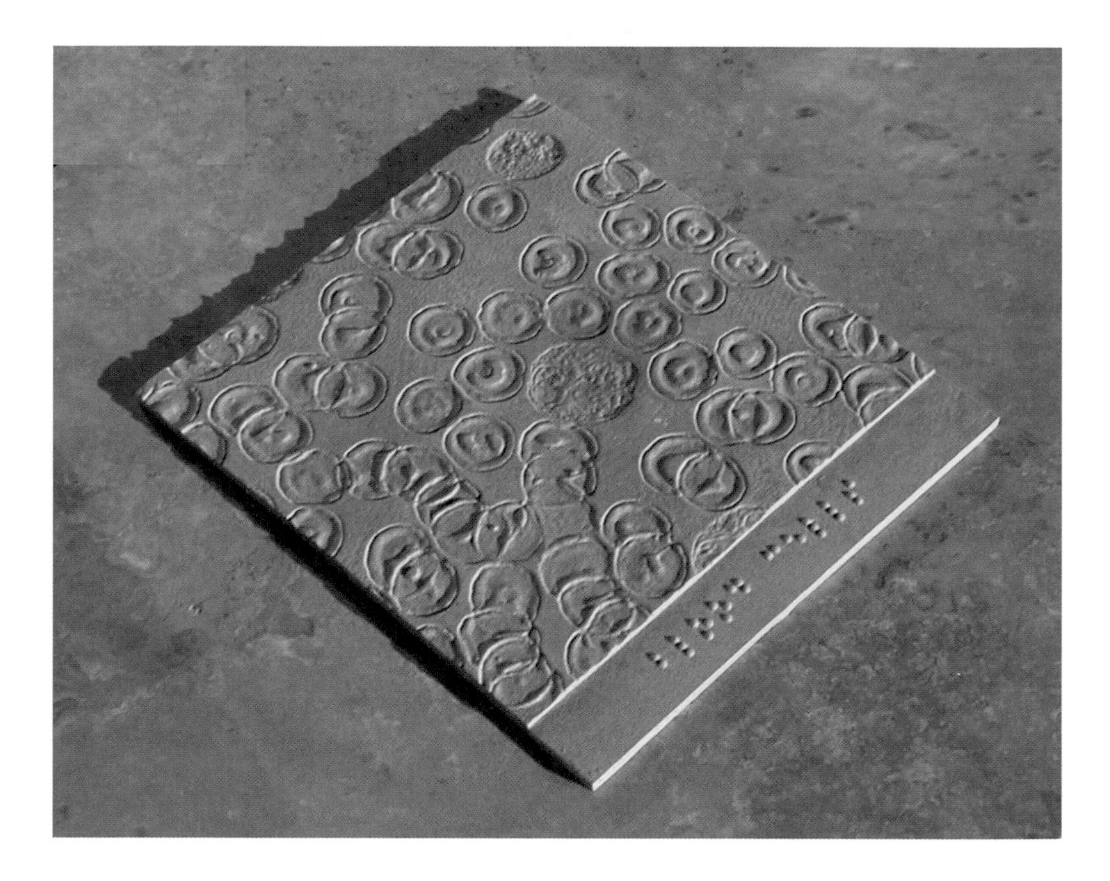

The inventor of this unique process is an architect, Keith Carlson. He intends to create a series of boxed sets of his tactile graphics. Each set will contain a number of tiles relating to a particular theme. For example, the Microscopic set will include 10–20 microscopic images, one of which may be of blood cells. The words 'RED BLOOD CELL' would be written in Braille and located at the bottom right-hand corner of the tile.

Other topics to be covered include Animals, Plants, Landscapes, Manmade Structures, Machines, Astronomy and Geography.

Each set will come in its own carrying case and will be available in libraries, schools, universities and organizations that serve the blind community.

During an initial round of user testing at a foundation for blind children, Carlson placed a tile featuring an image of Saturn on the table for one of the children to try out. As her fingers examined the surface, she asked 'What is this?' He told her to move her hand towards the bottom right-hand side to find the Braille description. As she read the word 'Saturn', she exclaimed, 'Oh – so that's what Saturn looks like!' When the entire room burst into laughter Carlson knew he had a successful project to work on.

**How sensitive are your fingers?** Your hand has about 17,000 touch receptors. These, together with nerve endings, send messages and information to your brain.

If you're walking in the dark, you instinctively stretch out your arms and fan out your fingers so your fingertips can detect what's in front of you. If your finger touches something painfully sharp its nerves will send a message to your brain, which in turn will rapidly send back the message to move your finger away from the source of pain.

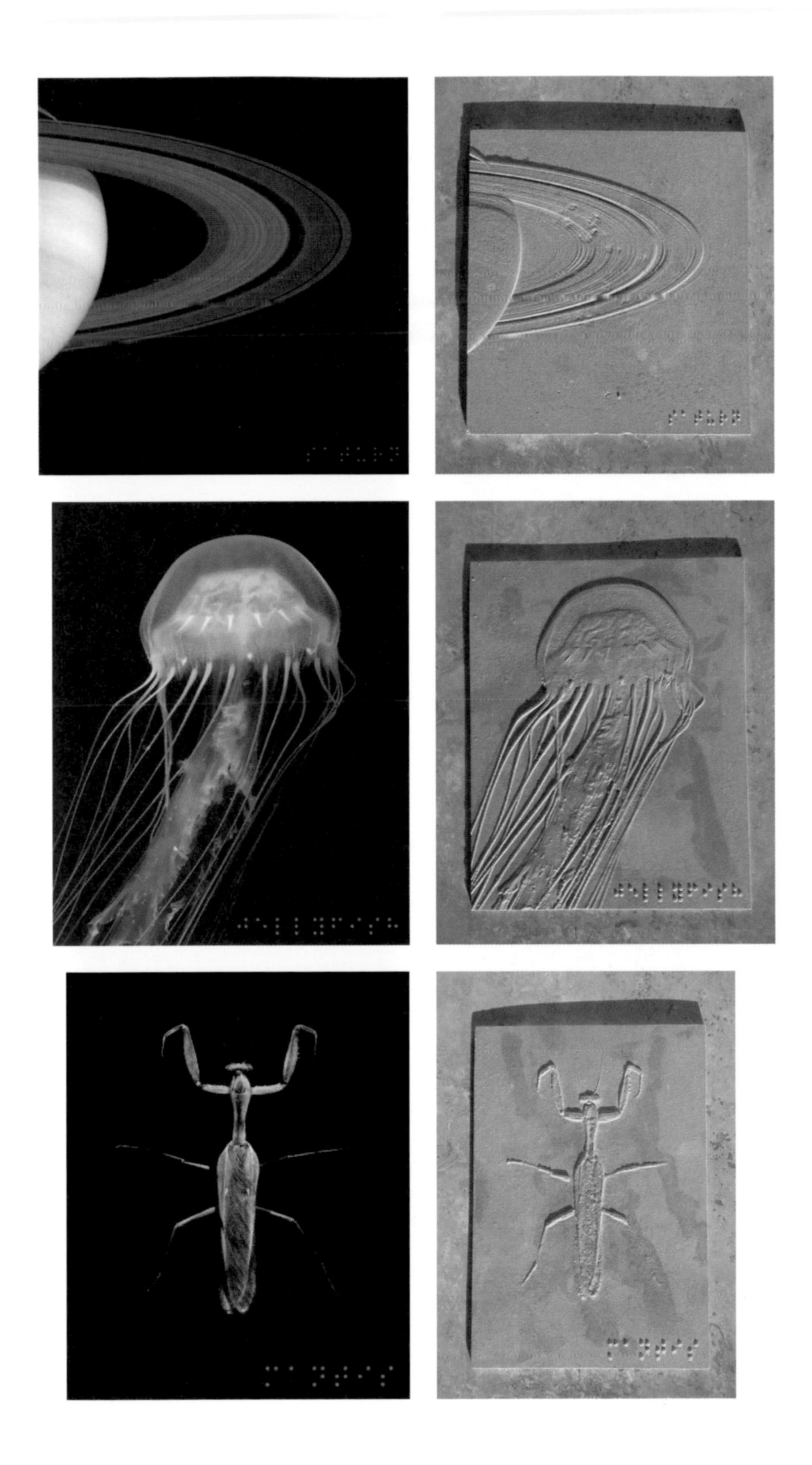

# How the blind can see with their ears.

KASPA
SAATCHI & SAATCHI AWARD WINNER, 1998

The Cold War, bats and the British Queen Mother all played a part in the development of Professor Leslie Kay's invention, KASPA.

KASPA stands for Kay's Advanced Spatial Perception Aid which, in principle, helps blind people to see with sound. KASPA emits ultrasonic sound waves. These produce echoes, which flow back to a sensor fitted on the blind user's head. The waves give the user information about the environment, such as the distance of objects, their direction, size, shape and even their texture.

KASPA's unique sensor elements convert the ultrasonic

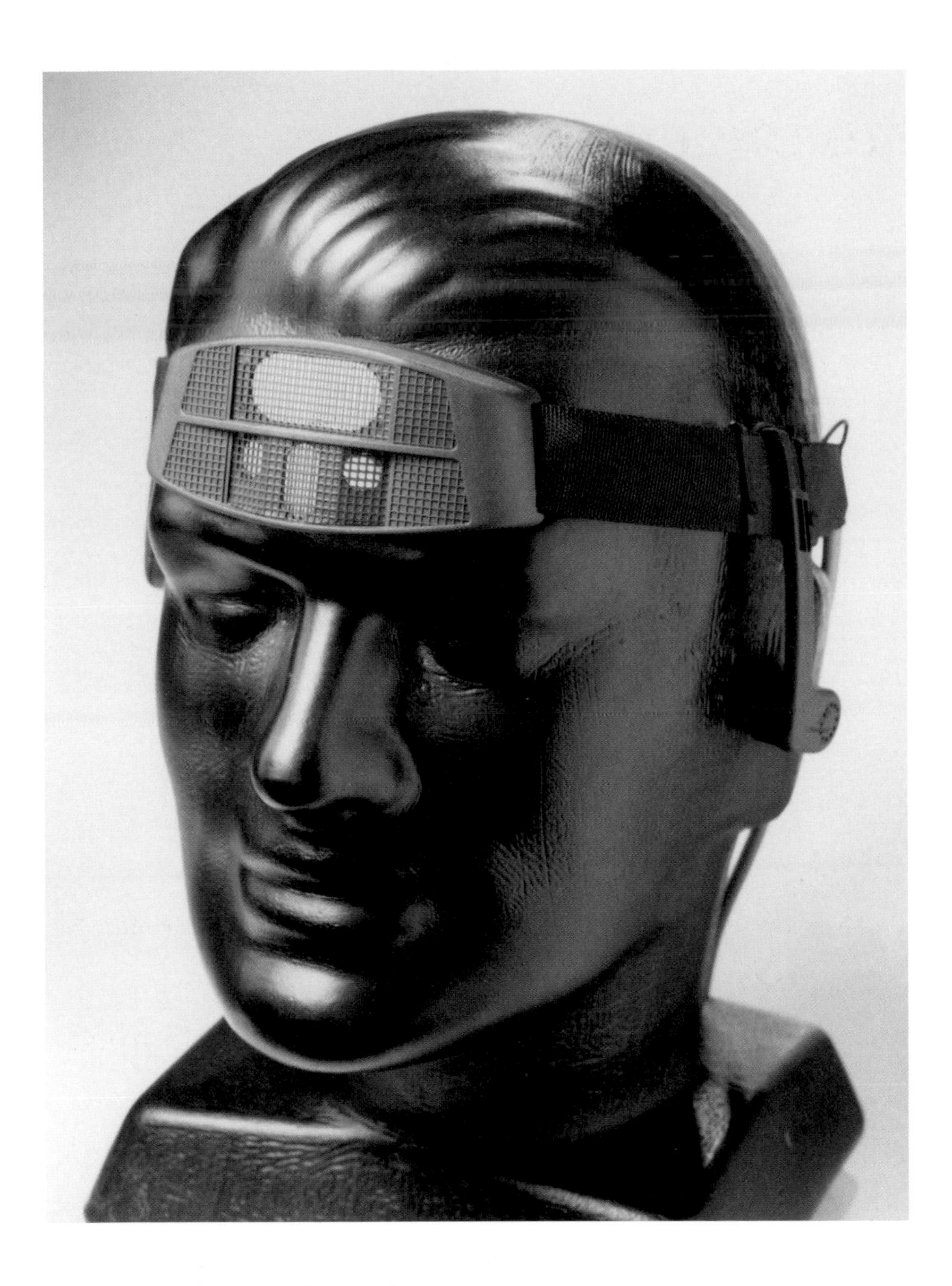

KASPA, the head-mounted sonar that models the dual functions of the eyes, providing peripheral wide-angle sonar vision and narrow central sonar vision up to 5 m (16 ft) distance.

KASPA
SAATCHI & SAATCHI AWARD WINNER, 1998

Card
SPECIALS
18
TODAYS SPECIALS

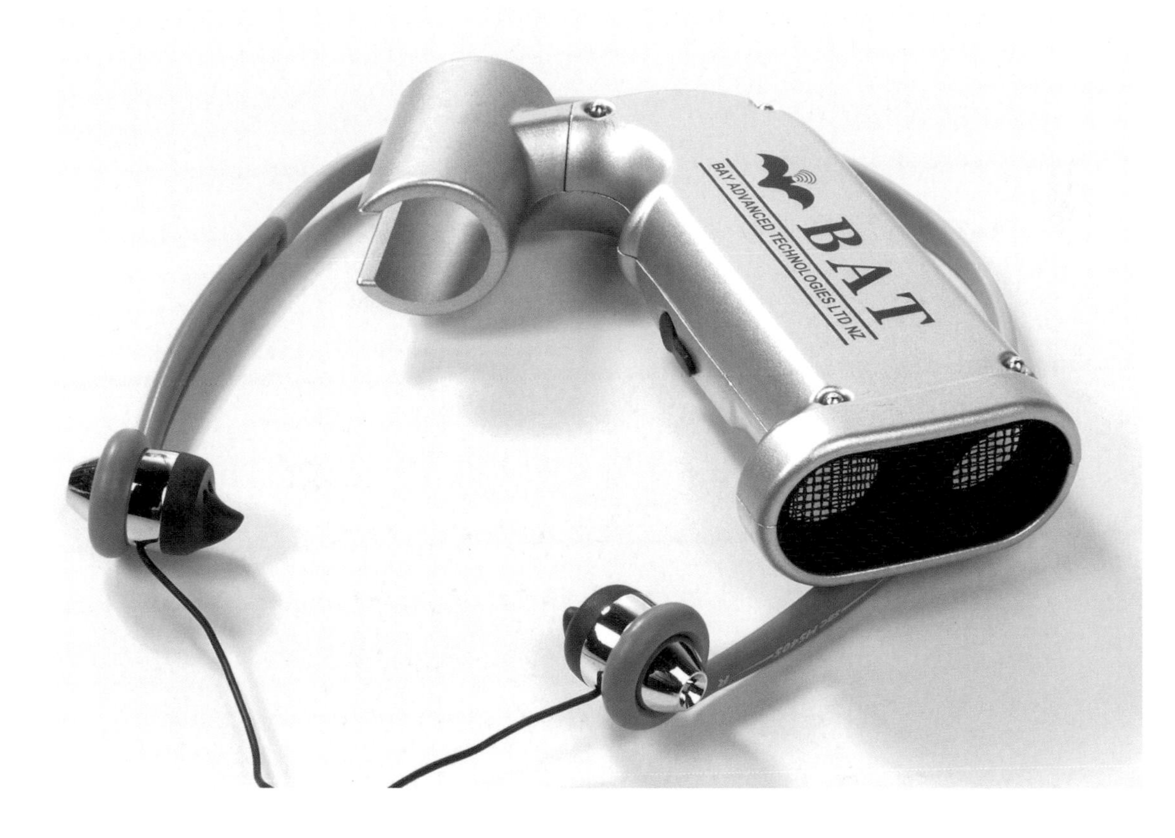


*Above* K-Sonar, an ultrasonic air sonar that slides on to a long white cane used by blind people.

*Left* Scenes from a demonstration video featuring a skilled KASPA user, Ivan Pivac.

waves into electrical signals, and these are processed to produce audible echo signatures, which carry the information, non-invasively, through miniature earphones, to the brain. Using neuronal auditory processing was a major scientific breakthrough, and the first of Kay's career.

The Sonic Torch was the predecessor of KASPA, and in 1965 Kay was the first person to receive Britain's National Scientific Achievement Award for the Sonic Torch as the first commercial use of ultrasonic echolocation in air.

The Sonic Torch emerged from Professor Kay's studies, with a colleague, Dr Griffin, of the sonar used by bats. The question they set out to answer was, 'Do bats use an octave band frequency sweep in their emissions so as to gain rich spatial information for environmental sensing?' The answer was 'Yes'.

The sonar that bats use is the same as the kind Professor Kay studied when he worked for the British Navy during the Cold War, developing underwater sonar technology that used sound waves and auditory neural processing to locate submarines, torpedoes, mines and so on.

KASPA, despite its breakthrough benefits, turned out to be a victim of economics, and is no longer on the market. But Professor Kay did not give up. He revisited the Sonic Torch, attaching it to a long white cane to create a two-senses aid for the blind, touch and sound. Professor Kay had the idea of combining the two in the 1970s, but miniaturization and other technological advances have only now made it a reality. The KASPA Sonar Cane is made by Professor Kay's own company, Bay Advanced Technologies Ltd which, of course, is BAT Ltd.

And the Queen Mother's contribution? She visited a major school for the blind in 1959, to open new facilities, including a swimming pool. This prompted Professor Kay, with his Navy experience, to wonder how the children would find their way about in the water and to set out to make them a sonic device. He hasn't stopped thinking about helping the blind to see with their ears ever since.

**How good is bat sonar?** We've been underestimating bats, apparently. Researchers at Brown University have discovered that a bat's brain can resolve sonar images up to three times more sharply than biologists had previously thought – and much better than manmade equipment.

Manmade sonar equipment can only process echo delays arriving five to 10 microseconds apart.

Bats, it seems, can resolve echoes that arrive just two microseconds apart as easily and routinely as if there were 10 microseconds between them.

A microsecond is just one-thousandth of a second. Couple this with a bat's ability to resolve echo-reflecting points on an object as fine as a pen line on paper, and you can see how finely tuned a bat's brain is.

# Getting the paralyzed to walk. Miracles no longer necessary.

The Stand Up And Walk Project
FINALIST, 2002

The name, The Stand Up And Walk Project, has a certain miraculous ring to it, especially for the paralyzed.

In Europe alone, this applies to around 300,000 people. Their average age is 31, and in 80 per cent of cases their paralysis is the result of traffic and sporting accidents.

Someone whose spinal cord is damaged at the neck will become quadriplegic and more or less lose the use of all four limbs. Damage on the thorax or small of the back may cause paraplegia, or paralysis of the lower limbs.

The medical reason is this. Neurons, the cells that make up the nervous system, are present from birth, but unlike other body cells they do not reproduce. Once destroyed, they cannot be replaced.

However, the muscles that lie beneath the site of the injury are still alive and are connected to the spinal cord by nerves.

The idea of restoring the mobility of paralyzed people came to Professor Pierre Rabischong 25 years ago.

Rabischong, of the Faculty of Medicine of Montpellier, France, and clinical consultant at Montpellier's Spinal Cord Injury Unit, started by using robotic solutions, to help paralyzed people by fitting active braces called 'walking machines'.

For technical and noise reasons these were abandoned, and Rabischong and his team started to focus on the electrical stimulation of nerves and muscles.

Rabischong and his team were finally ready to operate, and in 1999 their first paraplegic patient, Marc Merger, a Frenchman, had an electronic implant placed under the skin of his abdomen and connected by electrodes to muscles and nerves.

After a month's rest to allow the electrodes to become fixed, the training could start. The implant is controlled from outside the body by an antenna, which transfers power and a signal from a portable programmer.

There are three different programmes: standing up and sitting down, walking in a semi-automatic mode on flat ground, and voluntarily controlling the position of the foot.

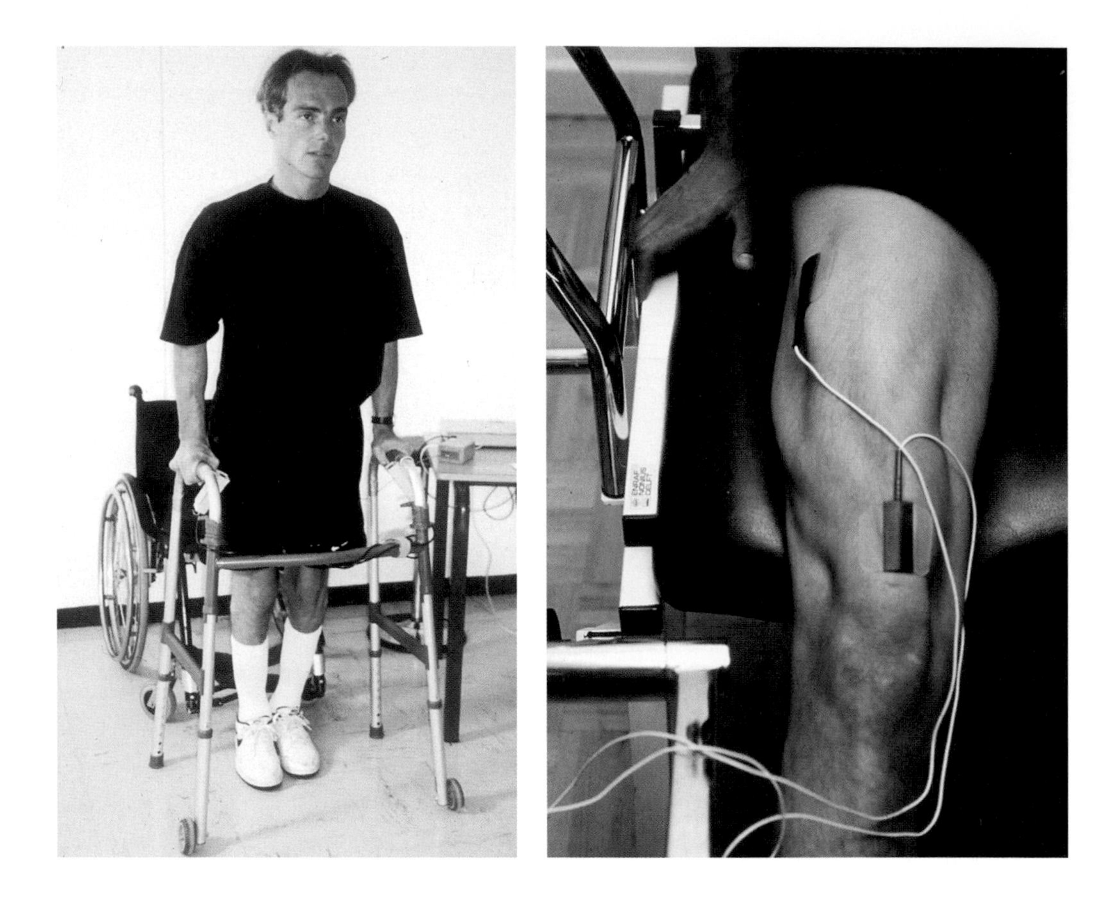

A push-button system allows patients to control their own movements.

Fully working, the implant allows a patient to actually stand up using their own muscles and without a brace, and to walk by pushing a button on the frame. The portable programmer, explains Professor Rabischong, 'does what the brain is normally doing by sending the right signal to the right muscle at the right time'. Which is another way to describe mobility.

**We are nervous** Your nervous system has two main parts, the central nervous system (CNS) and nerves. Your brain and spinal cord make up the CNS, which is in charge of operations.

Trillions of nerve cells, or neurons, in the CNS receive information that it analyzes and stores. They also send out instructions around your body.

Nerves are made up of bundles of nerve fibres, or axons. These relay impulses to and from all parts of your body.

# Control a computer. Look, no hands!

BioMuse
FINALIST, 1998

BioMuse is an example of an invention not necessarily inspired by the needs of people with disabilities, but that is certainly one of its valuable applications.

The starting point for Dr Benjamin Knapp, an electrical engineer at Stanford University, was to find a hands-free means of operating a computer. He figured, why take one hand off the keyboard to move the curser when your head is probably making the same tracking movements as you watch the screen?

The components of BioMuse are a lightweight headband, an armband (not unlike a futuristic-looking watch) and a base unit (a bioelectric signal controller). The headband allows you

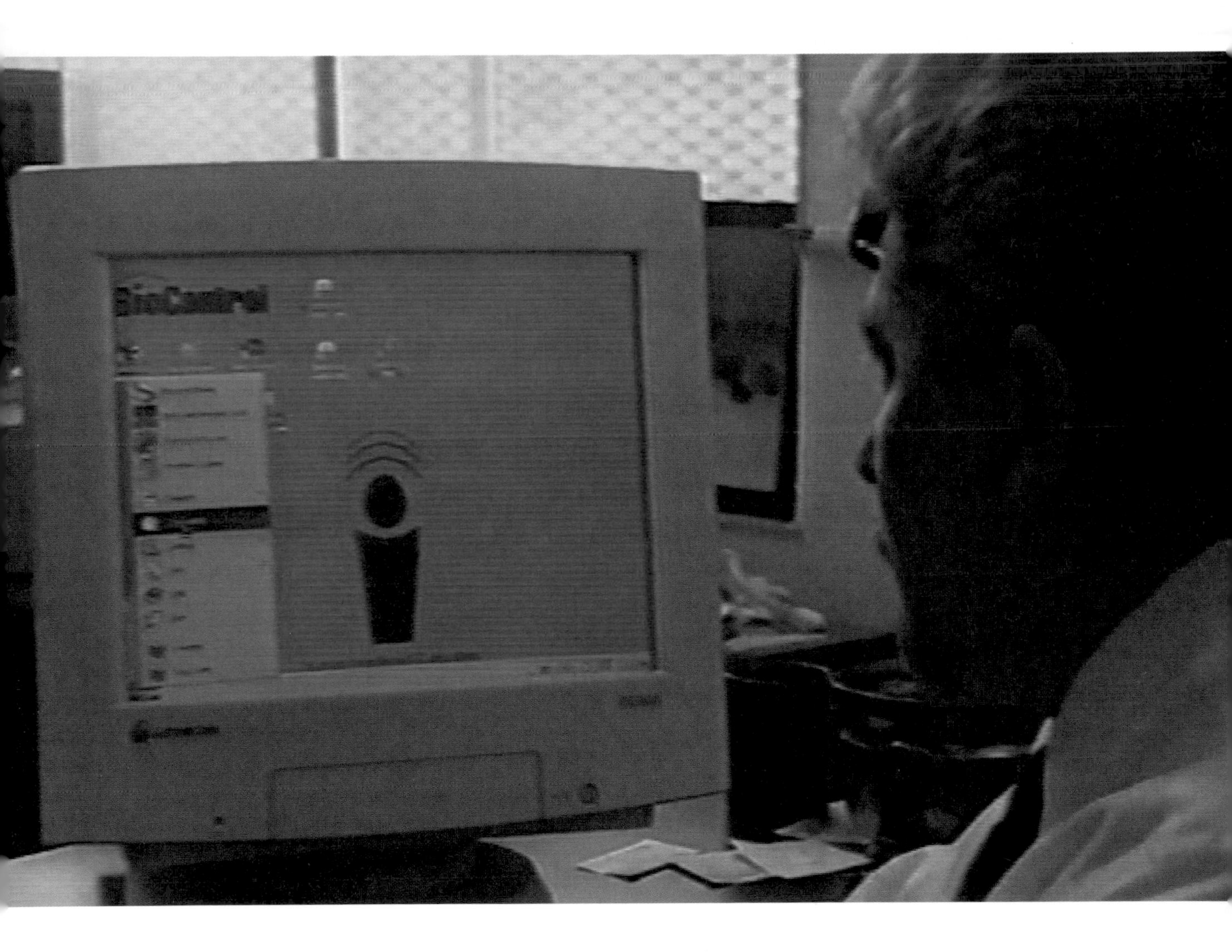
BioControl

to move the cursor as you move your head. Eye blinks activate button clicks. The armband tracks the position of your arm and monitors muscle tension.

The headband and the armband are standard non-invasive transdermal electrodes. They sit on the skin, in other words, and pick up electrical activity through it.

In all, the base unit receives data from four main sources of electrical activity in the human body. Muscles (EMG signals), eye movements (EOG signals), the heart (EKG signals), and brain waves (EEG signals). The base unit amplifies and digitizes these biosignals and processes the signals to control the computer.



BioMuse is a relatively low-cost set-up with numerous benefits, including reducing the risk of repetitive strain injury. It also gives people with certain disabilities access to a PC, perhaps for the first time.

**Your electric body** Electricity is involved in every cell in your body, and every second hundreds of tiny pulses of electricity travel through your body using your nerves as living wires.

These pulses are messengers, carrying information from the outside world to your brain, and instructions from your brain to the rest of your body. This two-way communication involves the nerve cells, or neurons, which can transmit electrical messages at high speed.

Each neuron has a number of inputs, called dendrites. Messages travel along these into the main cell. There's a single output from the cell called the axon.

Neurons do not touch each other. Messages cross the gap, or synapse, between neurons with the help of a chemical.

There are 100 billion neurons around your body and the same number in your brain.

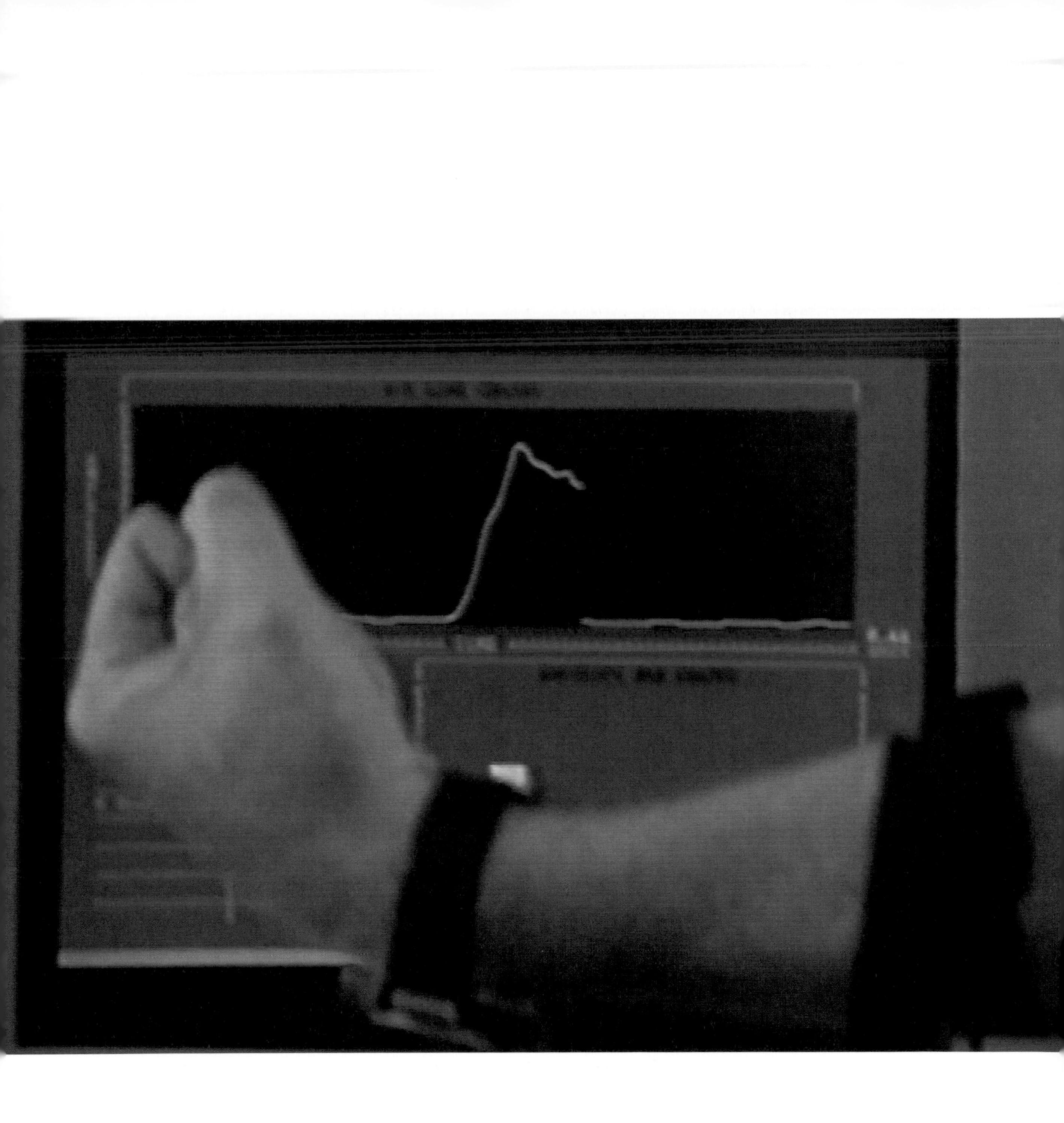

# At last. A Braille writer that writes the right way.

Jot a Dot
FINALIST, 2005

Braille was invented in 1829 by 15-year-old Parisian Louis Braille, who had been blind since the age of three, and it remains the foundation of written communication in education and employment for all blind people.

In 2002, the World Health Organization's World Population Statistics revealed that there are 42 million people in the world who are blind. Meanwhile, the progress in Braille writing has been barely at crawling pace.

The most common device still used for writing Braille was also invented by Louis Braille, in 1834. It's called the Slate and Stylus, and it requires the writer, having learned the Braille

UNO

code, to write the code backwards right to left, so that the embossed dot text is the right way round when it's read back.

The second most widely used Braille writing tool in the world is the Perkins Brailler, developed at the Perkins School for the Blind in Massachusetts in 1950. It looks like a mechanical typewriter from the period, and using it is so physically demanding that young blind children often have to wait until their arms are strong enough to start writing.

In contrast to these two devices, Jot a Dot is made of high-impact-resistant plastics and weighs less than 0.5 kg (1.1 lb), making it easy for anyone to carry and to write Braille anywhere.

Jot a Dot has a standard six-dot Braille keyboard, and enables writing from left to right. The six keys have been ergonomically designed to suit the widest possible hand size and shape.

Conveniently, it's also possible for writers to check what they've just written, and to start writing again from the same position they stopped.

Jot a Dot was developed by the Australian company, Quantum Technology (www.quantumtechnology.com.au) with initial funding from the Guide Dog Association of New South Wales and ACT.

**Braille** Louis Braille invented his reading system after attending a demonstration at the National Institute for Blind Youth in Paris.

The demonstration was of a code of dots and shapes embossed on card used by the French military for secret communications.

Louis adapted the complex code and the system he developed was standardized in 1932. It uses cells, each one a pattern of up to six raised dots in three rows of two. Sixty-three combinations are created using two sizes of dots.

Each of the cells represents a letter, a number, a punctuation mark, a common word, a common letter combination or speech sound, such as 'ah'.

There are now Braille adaptations for every major world language, and for music, mathematics and science.

# Change your own world by just thinking about it.

Mind Switch
FINALIST, 2002

Mind Switch gives even severely disabled people a level of genuine independence by allowing them to control electrical appliances through the power of their minds. Someone who can't even talk or move can change the channels and volume on their television.

The Mind Switch system was originated in 1995 and since then has been continually developed and improved. The latest manifestation has three surface electrodes incorporated in a cap worn by the user. Leads from the electrodes extend to a small box at the person's side. When the appropriate conscious signal arrives from the brain (which takes between

Mind Switch components: computer with Mind Switch software, RF receiver and transmitter, brainwave activity recorder.

The transmitter on the Mind Switch user's cap sends his brain signals to the controlling computer.

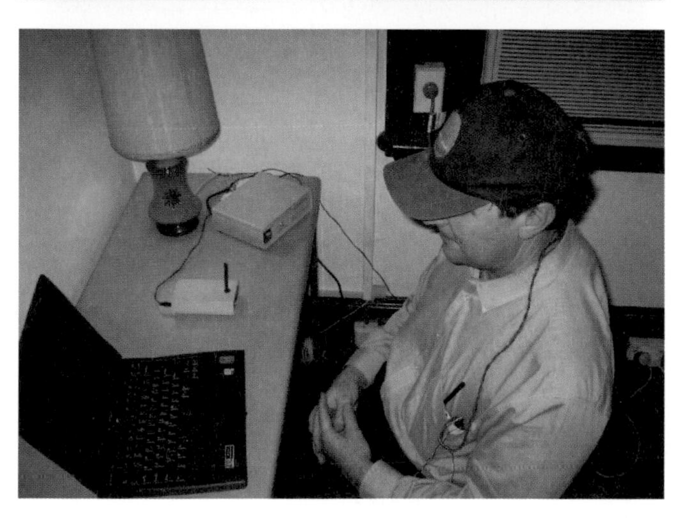

The user turns the lamp on and off using Mind Switch.

The electrodes that are placed on the user's head to measure the brain's alpha waves.



one and three seconds), the box sends an infrared or radio wave to the target appliance, which is then activated.

In a field trial, a patient turned on a television in around 10 seconds each time, and was able to turn the sound up or down and change channels within 20 to 30 seconds.

Dr Ashley Craig, who developed Mind Switch, describes the reaction of the patient: 'While these times are obviously longer than able-bodied people would endure, it was extremely exciting to her that she had the ability to control her TV independently with reliability.'

Dr Craig is Professor of Behavioural Sciences in the Department of Health Sciences at Sydney's University of Technology. He explains that although 'assistive devices' are not new, few can be used by the disabled. 'The weak link in most control systems is the method used for switching, that is, the type of interface between the system and the person.'

Typical methods are 'suck-puff' techniques, chin-operated control sticks, mouth-sticks, voice control and eye blinks. 'Simple switches such as a chin-press have been used successfully. However, this assumes a person has control over their neck and head movement.'

Many victims of high-break spinal injury (quadriplegia), a massive stroke, cerebral palsy, multiple sclerosis, polio, muscular atrophy and advanced motor neurone disease are so disabled they can't use traditional control systems such as voice-activation (up to 40 per cent of the profoundly disabled can't speak) or systems that require some use of the arms.

Other scientists have attempted to use brain signals to control electrical appliances like computers or TV but few, if any, have recorded efficiency rates above 50 to 70 per cent. Users of Mind Switch have achieved 95 per cent efficiency levels, with little or no training. All they need to be able to do is relax sufficiently to generate the alpha brain waves that trigger the control.



**Brain waves** The human brain contains around 100 billion neurons, or nerve cells. These send electrical signals, which can be detected by electrodes placed on the scalp when using an electroencephalogram (EEG) machine. The voltage is tiny (around 10 microvolts), but when it's amplified and traced out on paper, a wave pattern emerges. There are a number of different patterns. An alpha wave appears when you're relaxed and alert. They're best detected with the eyes closed. A beta wave is faster and is associated with active, busy or anxious thinking. Delta and theta waves are slower and are linked with stages of sleep.

# Blind person sees after 50 years!

Biochip Artificial Vision Project
FINALIST, 1998

In a healthy eye, the 130 million or so cells of the retina collect light and images, and provide you with your pictures of the world. But age-related degeneration of the retina, for example, gradually reduces this function. Sufferers may live for many years but have to manage with little or no sight.

Now there's progress in the shape of computer microchips.

Two things motivated Dr Mark Humayun's Biochip Artificial Vision Project. One was his grandmother, who had helped raise him, losing her sight through diabetes. The other was facing the fact that even patients treated at the world's best eye institutions continued to go blind.

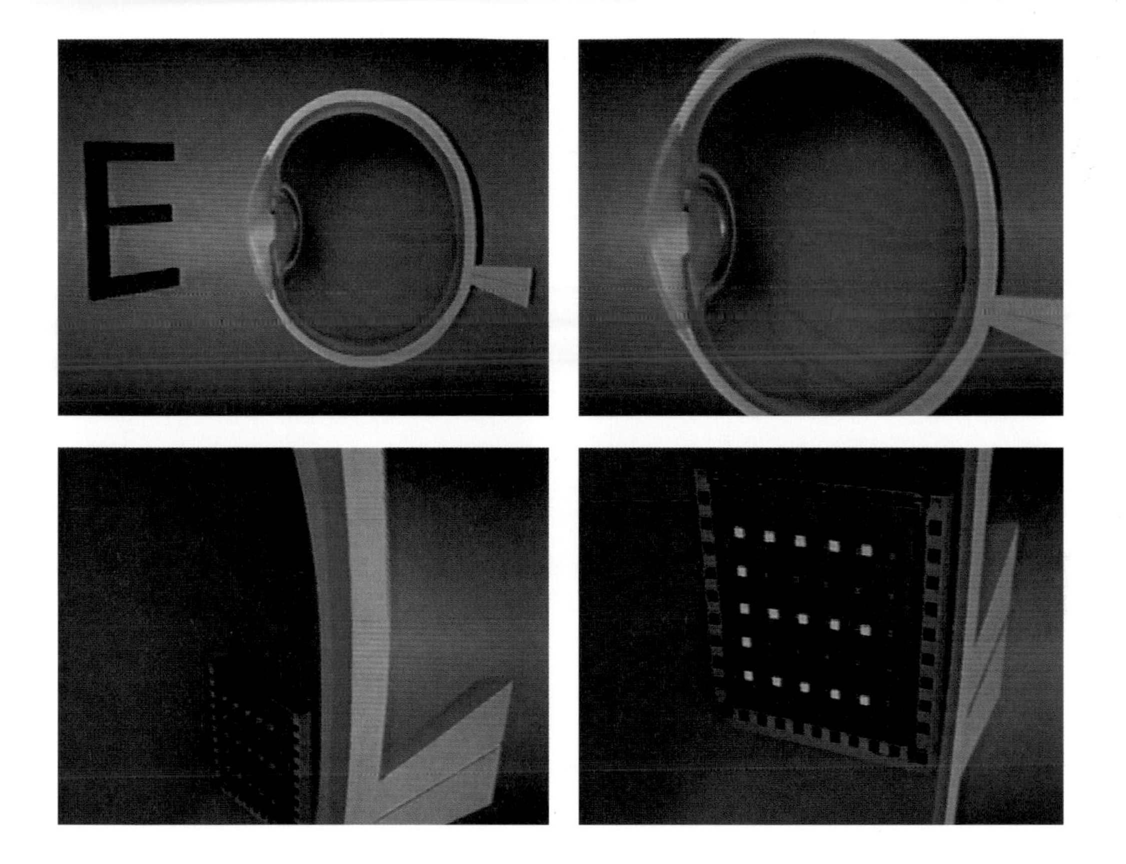



Dr Humayun teamed up with Dr Eugene de Juan with the idea of producing a retinal microelectronic implant.

Their Biochip Artificial Vision Project is the outcome.

It involves surgically attaching a flexible silicon biochip near the retina. An external camera mounted in an eyeglass frame captures an image and converts it into an electrical signal that is transmitted to the biochip. The chip electronically stimulates the healthy retinal cells that send the signals conveying the image to the brain. The image is made up of a pattern of light dots similar, in effect, to a dot-matrix printer.

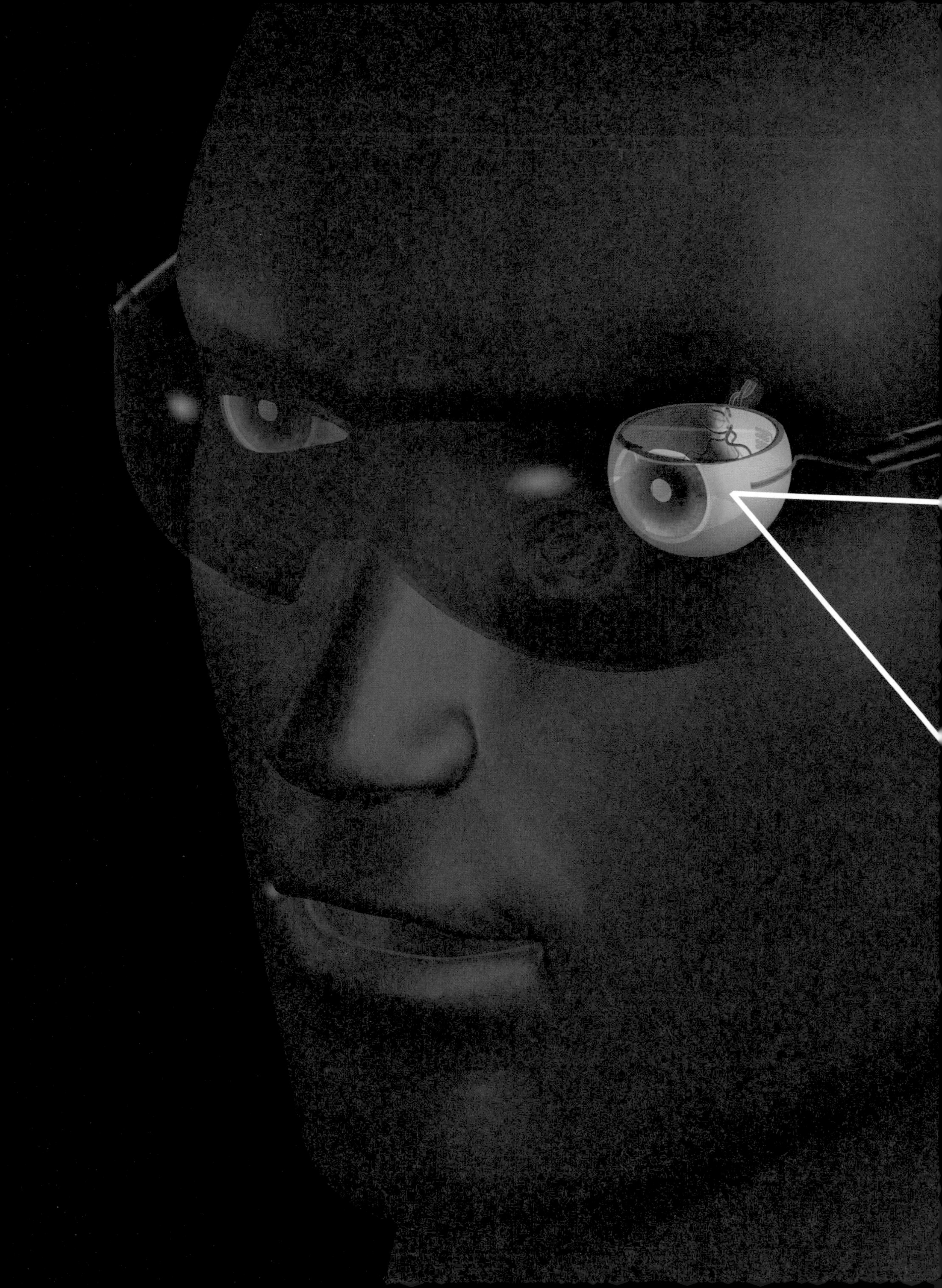

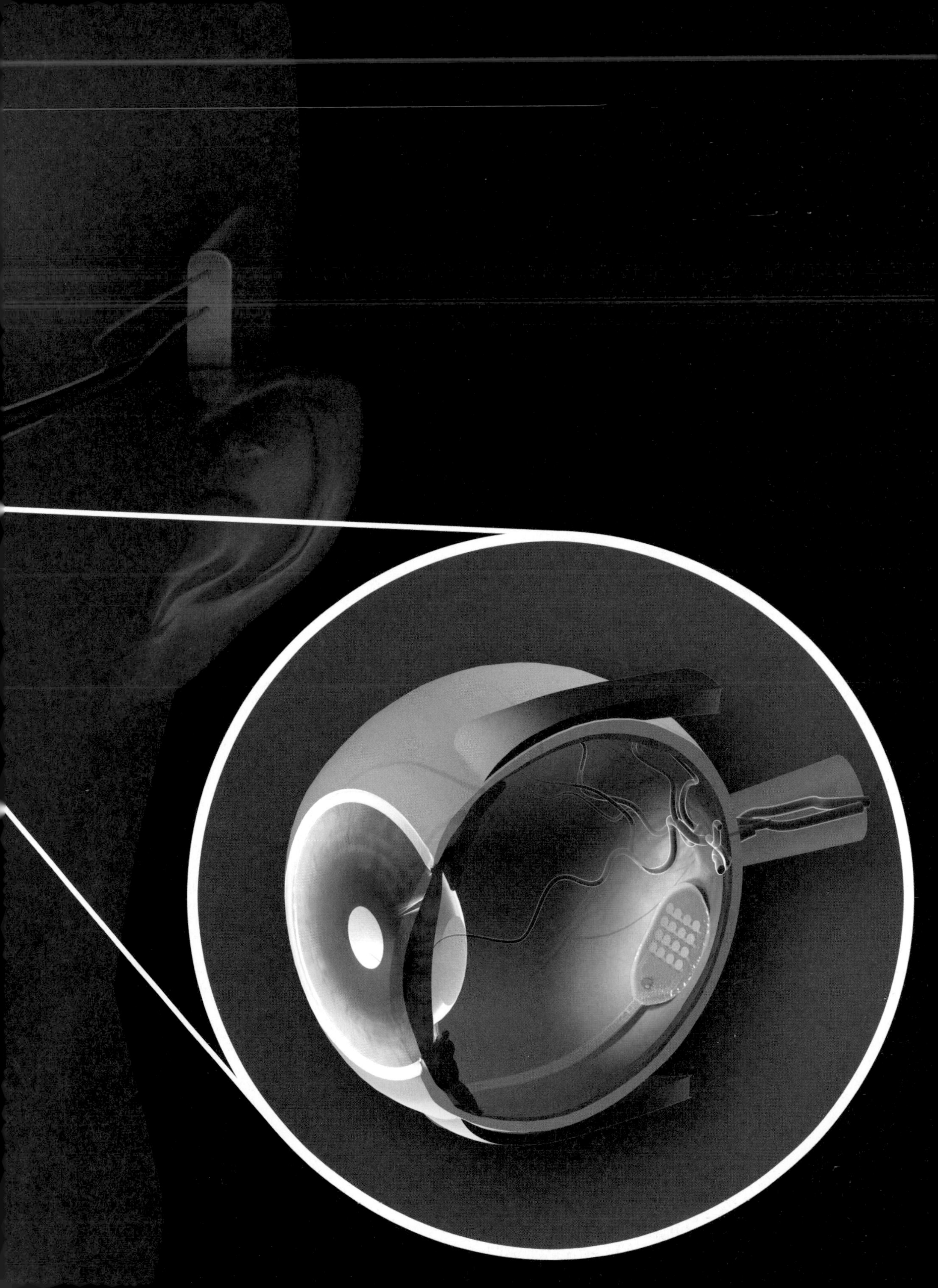

Dr Humayun implanted the first biochip after 15 years of research. The patient had been blind for more than half a century, but the results were astounding. The patient was able to recognize large objects such as chairs, and to differentiate between a cup, a plate and a knife. The first implanted biochip was still working more than four years later.

Dr Humayun is Professor of Ophthalmology, Bio-Medical Engineering and Cell and Neurobiology at the Doheny Eye Institute, University of Southern California.

His team at the Artificial Vision Project wants to develop an implant that makes a more detailed image possible by increasing the number of pixels. Ultimately, this might lead to blind people picking up a book and reading again.



**How your retina works** Your retina is a membrane, packed with light-detecting photoreceptors called rods and cones. The retina's five to seven million cone cells detect colours, and its 125 million rod cells work in dim light and see in black and white.

There's a small yellow area at the centre of the retina called the fovea. Here, each cell has its own connection to your brain, and is the most sensitive part of the retina.

When what you see reaches your retina it's upside down, but your brain turns the image the right way up.

# Blind man drives car!



Work on the Artificial Vision for the Blind system began in 1968, but the project has gathered momentum as computer technology has advanced.

The man with the vision was the late Dr William Dobelle, a researcher with Batchelor and Masters degrees in Biophysics and a PhD in Physiology. Dr Dobelle focused his career on improving the lives of people who could benefit from implanted medical devices.

The Artificial Vision system works for almost all causes of blindness in people, apart from the two per cent who are born blind or become blind while children (the visual cortex may take between eight and 10 years to fully develop).

BIONIC EYE
CNN
USES DIGITAL CAMERA
AND COMPUTER
BIONIC EYE
CNN
USES DIGITAL CAMERA
AND COMPUTER

BIONIC EYE
CNN
BIONIC EYE
CNN

BIONIC EYE
CNN
BIONIC EYE
CNN

BIONIC EYE
CNN
BIONIC EYE
CNN

Patients use special sunglasses fitted with a miniature television camera. Information from the camera is fed to a microcomputer and then to a stimulator, both worn on a belt around the patient's waist.

The stimulator is connected to two implants touching the brain via two cables attached to two tiny fire-hydrant-like connectors mounted in the back of the skull. The visual cortex provides a two-dimensional map of visual space and is found at the back of the brain.

A small spot of light can be created for the patient by stimulating the visual cortex with tiny, surgically implanted electrodes. When a patient is stimulated with a number of these electrodes, the visual effect resembles a sports stadium scoreboard.

This doesn't give patients 'normal' vision. They see white dots of light (or phosphenes) that resemble stars on a black background and they learn to interpret the patterns. In effect, major objects show up in outline, with visual detail omitted.

Between April 2002 and January 2004, 16 patients were implanted. Fifteen could see the dots of light. After training, several could walk in the street, although using a cane. Their independence, mobility and self-esteem were greatly increased. The one patient who could not see the dots of light had lost his sight when he was four, and both he and his parent were aware that the possibility of seeing was extremely remote. Two patients who had been totally blind have learned to use the system well enough to slowly drive cars.

Dr Dobelle lived long enough to witness this success and to declare that 'it's amazing now. It was science fiction a few years ago.'

When he died in October 2004, the project and his patent were donated to Stony Brook University in Long Island, New York.

The Dobelle Institute no longer exists, but the belief lives on that Braille, the white cane and the guide dog will be made redundant by their system by the end of the century.

**A look at the visual cortex** A visual cortex is a sight centre that decodes and analyzes nerve signals from cells in the retina.

A visual cortex is made up of regions, each of which has a number. The primary visual cortex, which is the main reception area for visual signals, is V1. Around this, in the secondary visual cortex, are regions numbered V2 and upwards.

Each region sorts and, to some extent, separately processes the various aspects of vision, including colour, contrast, distance and depth, shape and form, and movement.

V4, for example, seems to be engaged principally with the analysis and comparison of colours and contrasts.

The results of this parallel processing are recombined as these regions communicate in the other parts of the brain, to complete the picture of what we're seeing, including names and meaning.

## THE AWARD FOR WORLD CHANGING IDEAS

In 2005, we decided to change the name of the Award to better reflect our inspirational dream – 'to be revered as a hothouse for world changing ideas.'

Over the years the prize-giving event has been hosted in a variety of venues around New York.

In 2006, we were able to do it in the huge and newly renovated gallery space in our Hudson Street headquarters.

As the Award has grown, so has its ability to attract an A-list of innovative thinkers to the event.

Amongst them, it was great to have this time, two judges from the very first Award, Edward de Bono and Laurie Anderson.

Laurie came with Lou Reed who had been on the most recent panel.

Along with the other finalists, Jimmy Wales, the founder of Wikipedia, was also there.

"Wikipedia represents the best and the worst of the internet," Lou Reed told me.

"The best for all the wonderful and obvious things about the internet, and the worst because the information's not always accurate. For example, they misspelt my name, now it's out there ... Yesterday I had to argue with a journalist on the spelling – I offered to send the guy a copy of my birth certificate!"

When Jimmy Wales heard about Lou's story, he was very amused and said he couldn't wait to tell his staff!

7

# The Developing World

# You can be your own optician.

If you live in the developed world and you can't see properly, you go to an optician. But for the majority of the estimated 500 million people in the developing world who can't see clearly, this isn't possible, either because there are no eye-care professionals or because the cost is prohibitive.

Professor Joshua Silver's Self-Adjustable Spectacles overcome these twin hurdles. His invention is simple and inexpensive, and the wearer becomes his or her own eye-care specialist.

How do DIY, Self-Adjustable Spectacles work? Their lenses are filled with fluid and the power of the lenses is changed by

varying the amount of fluid in each lens until the clearest vision is achieved. This is done manually and takes less than a minute for both lenses.

Once the optimum vision is found, the lenses are 'set' and sealed, and then the adjusting devices, two small pumps with short lengths of tubing, are taken off.

A Professor of Physics at Oxford University, Joshua Silver is an experimental physicist, and now he's also the chairman of the Adaptive Eyecare Company.

Their production plant in China manufactures 2,000 pairs a day. The company has contracts to supply 46,000 pairs and a pipeline of orders that runs into millions, which is hardly surprising, given the size of the potential market.

Professor Silver does not claim the invention all for himself. Fluid-filled, variable-power lenses were the idea of a German scientific instrument maker in 1747, and there were patents for such spectacles throughout the twentieth century. As the Professor says: 'Often a device is evolutionary because of the nature of invention. There are ideas whose time has come.'

So far, the spectacles have been sold in Africa, because the need is most prevalent there. In fact, the first pair that was ever used in the field, in Ghana, is now on display in London's Science Museum.

– +
6543210123456

As Professor Silver observes, it is unusual for new technology to be applied to the developing world, but there are many people in Africa already who must be very glad this particular technology has been.

**What spectacles correct** Spectacles, glasses or eyeglasses are used to correct three principal focus problems affecting sight.

Myopia (short-sightedness or near-sightedness) is a condition where near objects are in focus and more distant objects blurred.

Hyperopia (far-sightedness or long-sightedness) is the opposite. Near objects are blurred and distant objects are in focus.

Astigmatism is caused by the cornea, or lens, of the eye not being perfectly spherical and results in a loss of visual detail.

In myopia, parallel rays of light entering the eye converge before they reach the retina.

In hyperopia, they don't converge before they reach the retina.

In astigmatism, the rays converge at two separate locations either before and/or after the retina.

Emmetropia is the condition you don't need glasses to correct. It describes ideal focus, where the parallel rays of light converge on or at the retina.

# Turning a single mud wall into a reference library.

Kinkajou Projector
FINALIST, 2002

Twenty per cent of the world's adult population can't read. Books are expensive to print and ship to the developing world, and attempts to improve literacy levels often revolve around badly lit evening classes. In this rather depressing context, the Kinkajou Projector is something of a beacon of light.

The Kinkajou Microfilm Projection System, to give the Projector its full name, is the invention of Design that Matters (DtM) in Cambridge, Massachusetts. DtM focus on under-served communities and all their work is carried out by volunteers.

The Kinkajou Projector is a low-cost teaching tool, designed to improve and expand access to education by

sayi yɛrɛ yada
kalaya
teriya
ja a
ji i
jija i jija

transforming night-time learning environments in rural areas without electricity.

The Projector is an innovative combination of cutting-edge hardware, creative re-purposing of existing products and 'abandoned' technology.

It combines Light-Emitting Diode (LED) technology with microfilm, and an adaptation of the plastic lenses used in View-Master toys. Timothy Prestero of DtM describes the combination as 'bizarre'. It would be equally accurate to describe the apparent incongruity as the very essence of invention.

The five-watt white LED light source is rugged and rated to last 100,000 hours, or the equivalent of 11 years' non-stop operation.

Microfilm is durable and cheap. A reel of 10,000 images, enough for an entire reference library, costs just £3.20 ($6).

The Projector's optics assembly incorporates seven plastic lenses and the resulting system can cast an image from the microfilm up to 3 m (10 ft) wide on to practically any flat surface. This is big enough for an entire classroom to read.

The Projector could be adapted to run on any power source, including human power, but it currently runs on batteries charged by solar panels.

DtM's client for the Kinkajou Projector is World Education, which administers adult-education programmes in 27 countries in Africa, Asia and South America. So far, Kinkajou

Projectors have been working in 45 villages in Mali, helping over 3,000 adults learn to read. In 2007, the project will increase its coverage to 1,500 villages in West Africa, India and Bangladesh, reaching 500,000 illiterate adults over five years.

With production volumes of 10,000, the entire Kinkajou Projection System, including battery, charge controller and solar panel, costs just £64 ($120).

When it's compared with the cost of buying and shipping books, the Kinkajou Projector will cut the cost of education – long term – by 60 per cent which, in turn, means more people can be taught to read.

It's said that knowledge is power. Education is certainly one of the most powerful weapons in the war against poverty in the developing world.



**Literacy facts** Around the world, 130 million children aged six to 11 are not in school, and 90 million of these are girls.

More than half of the 130 million live in India, Pakistan, Bangladesh, Ethiopia and Nigeria.

A further 150 million children leave school without basic literacy or numeracy skills. One in four adults in the developing world is illiterate.

More than 50 per cent of women in Sub-Saharan Africa are illiterate. Over 80 per cent of women are illiterate in Burkina Faso, Sierra Leone, Nepal, Somalia and Afghanistan.

World Bank research shows that investment in girls is the single most valuable development intervention any country can make.

The illiteracy rate in Sierra Leone and Liberia is 80 per cent. They rank with Angola at the bottom of the UN's human-development index.

# Be heard miles away. Without a phone. Without shouting.

Critical Mass Communicator
FINALIST, 1998

How do you communicate across distances if you don't have a phone, you don't have electricity, you don't know how to read and you don't know how to write? This is the reality for vast numbers of people in the developing world.

Jaron Lanier has been described as a maverick genius, but he's also a computer scientist, composer, visual artist and author. He grew up in New Mexico. He's been credited with originating the term 'virtual reality', and he set up the first company to sell virtual reality products, VPL Research, in the early 1980s.

And in 1998 Lanier had an actual idea to change the reality described above.

He called it the Critical Mass Communicator – a human-powered, wireless, cellular device with speech recognition and translation capabilities.

An illiterate person would be able to 'dial' a CMC using their voice. The microprocessors inside the phone would decode the information the user speaks into it, such as thc recipient's name and location, and the message itself.

Then, instead of searching out the usual (expensive) cell systems or satellites, the CMC would pick up other CMCs nearby that were being cranked at the same time.

'Packets' of information that had been stored in one CMC would eventually be passed on to the intended recipient, via other CMCs that had picked up the 'packet'.

As Lanier explained, 'It's something that only works because of the collective use of the device but it depends on nothing but the collective use of the device'.

The CMC was probably the first communications tool that would require no infrastructure in its deployment area, and as it wouldn't need power, training or facilities, it would be impossible to regulate too. This would have been particularly appealing to poor people living under oppressive, censorious regimes.

Eventually the CMC might also have been programmed to share photographs and web pages, but at its basic it could have provided two fundamentals we take for granted in the

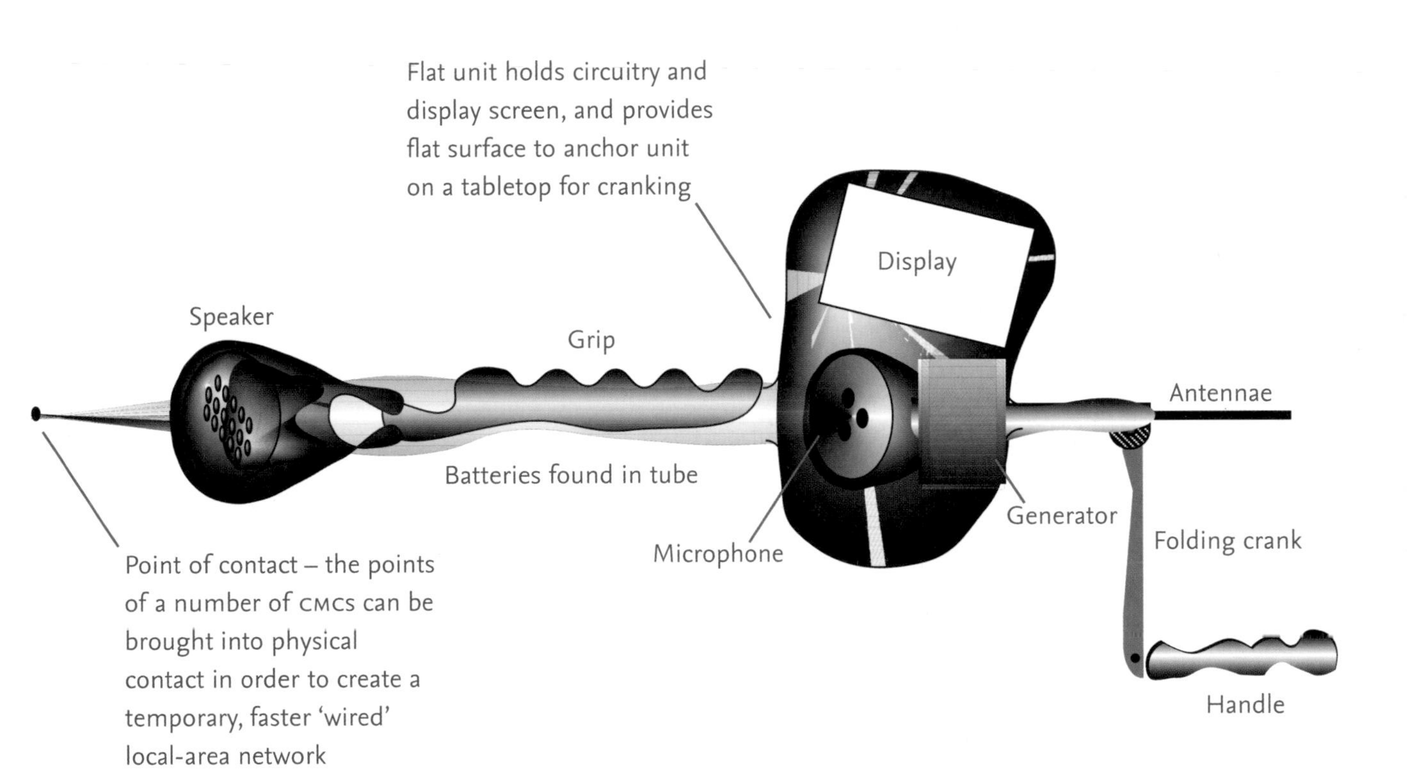
Flat unit holds circuitry and display screen, and provides flat surface to anchor unit on a tabletop for cranking
Display
Speaker
Grip
Antennae
Batteries found in tube
Generator
Folding crank
Microphone
Point of contact – the points of a number of CMCs can be brought into physical contact in order to create a temporary, faster 'wired' local-area network
Handle

developed world: knowledge and having a voice.

Technological advances since 1998 have made the CMC impressively prescient. One only has to consider the advent of peer-to-peer networking, for example, or Nicholas Negroponte's 'one laptop per child' initiative (the Hundred Dollar Laptop) to see how far ahead of the curve Jaron Lanier's thinking was.



**Virtual reality** In the early 1960s, a cinematographer called Morton Heilig had an idea. 'I became very excited. I thought, "why stop at a picture that fills only 18 per cent of the spectator's visual field, and a two-dimensional picture at that? Why not make it a three-dimensional image that fills 100 per cent of the spectator's visual field, accompanied by stereophonic sound?"'

Heilig's idea stalled through lack of financial support, but by 1969, Ivan Sutherland had created the first virtual world, as seen in a head-mounted display. His development of computer graphics was key to this.

'Virtual Reality', invented by Jaron Lanier, refers to a social or multi-person virtual world, making it, in essence, a clone of physical reality.

Since Sutherland, the uses of virtual world technology have proliferated, and include flight simulation and telepresence and teleoperating.

# A new optical device for seeing smaller bills.



You could say that there's more to sight than meets the eye. By examining the back of the eye it's possible to identify abnormalities that may indicate serious medical conditions. These may be sight-threatening diseases, like glaucoma, or life-threatening like diabetes, high blood pressure or cerebral malaria. The instrument used for 'retinal exams' is an ophthalmoscope (direct or indirect) and, traditionally, these are used by specialist ophthalmologists or opticians. Non-specialists like general practitioners, nurses, paramedics and other trained healthcare professionals may use the direct ophthalmoscope, but two main reasons prevent them.

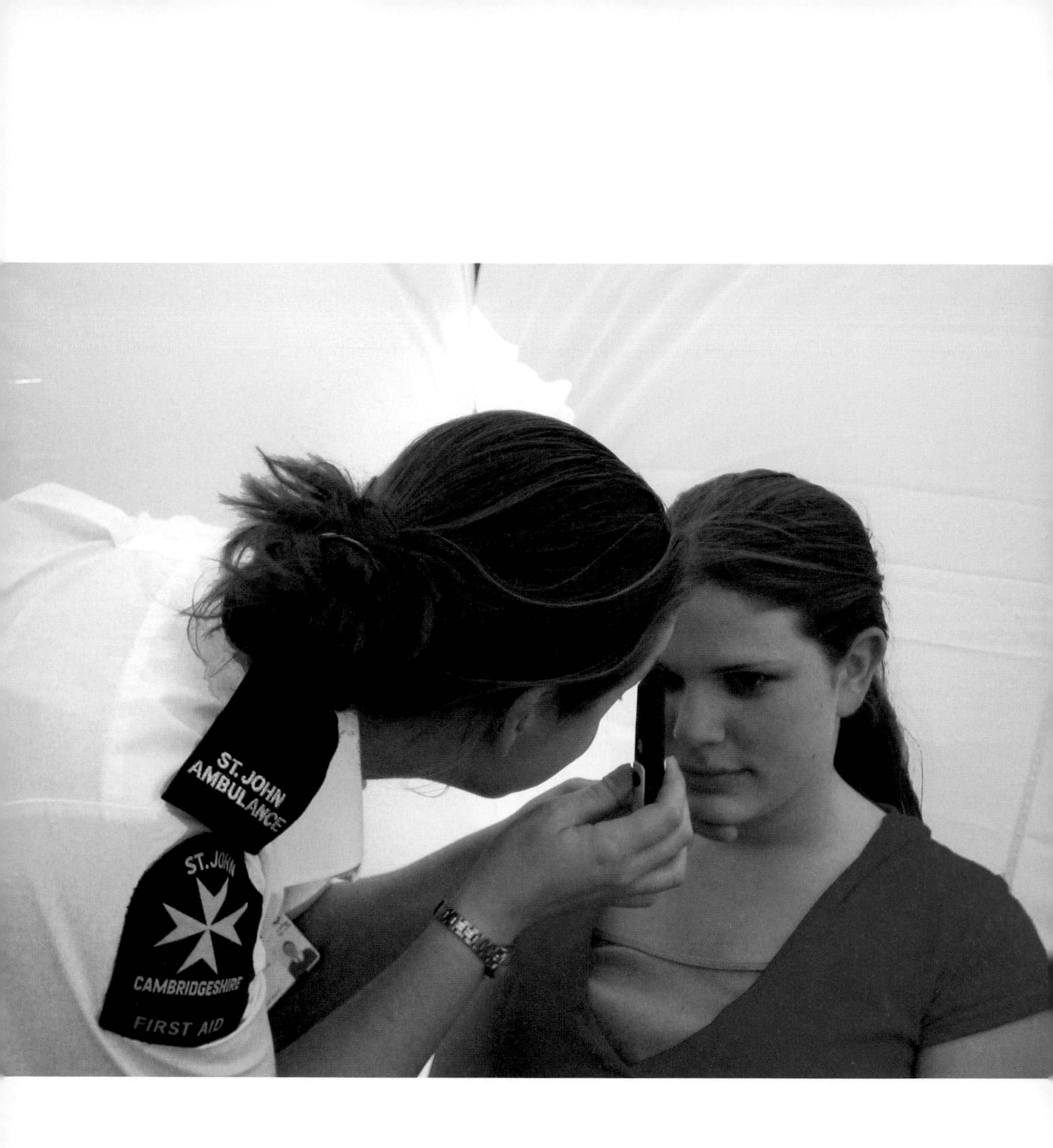
ST. JOHN
AMBULANCE
ST. JOHN
CAMBRIDGESHIRE
FIRST AID

First, the cost of a traditional ophthalmoscope is high (up to £1,000/$1,800), which many medical students and healthcare professionals are unwilling or unable to pay in order to have their own instruments.

In the developing world, this cost is prohibitive, not just for individuals but for general medical wards and clinics, too.

Second, current ophthalmoscopes have up to 40 viewing lenses which, together with apertures, have to be manipulated during the exam, making the procedure very complicated.

Because they don't tend to have an ophthalmoscope of their own, non-specialists have limited opportunity to perform retinal exams, and skills acquired during training diminish, as does confidence.

Roger Armour is a retired surgeon and is acutely aware of these issues. But his belief in the retinal exam as a powerful and revealing technique is so strong that he has invented Optyse, a new Lens-Free Ophthalmoscope (LFO).

His invention simplifies the technology of existing ophthalmoscopes by removing their lenses and apertures, but the performance of the Optyse LFO matches standard ophthalmoscopes. As a consequence, successful retinal exams could be carried out by non-specialist healthcare workers, particularly in the developing world.

With a list price of £38 ($68), the Optyse LFO also eliminates the cost issue, and opens up the possibility

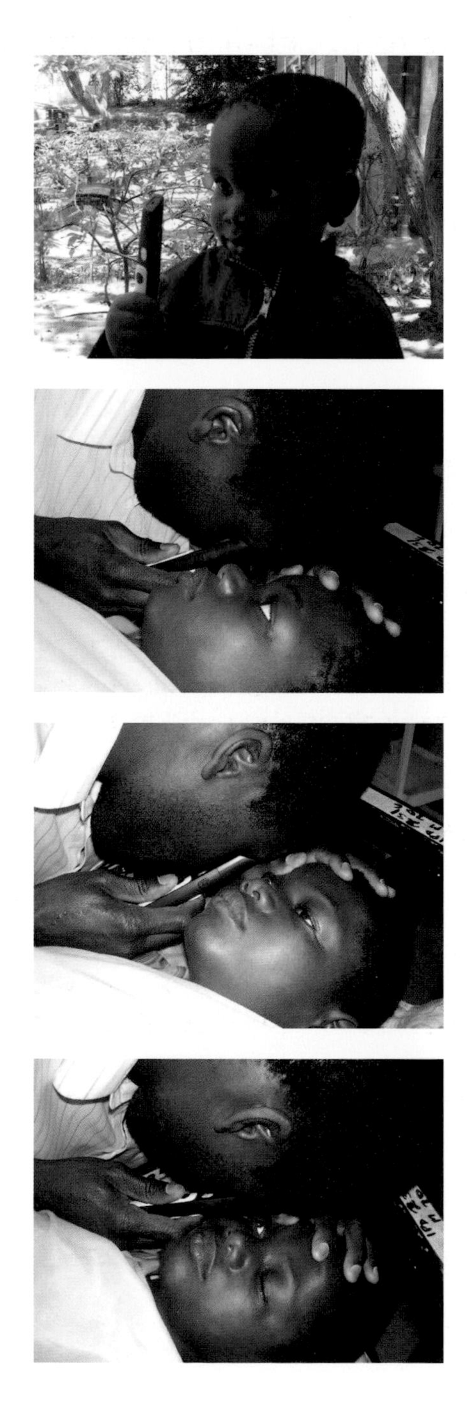

of tackling diseases such as hypertension and diabetes, which are emerging problems in Sub-Saharan Africa.

It's estimated that there are up to 42 million blind people worldwide, and that 75 per cent of these could have their sight restored or are blind from preventable causes. The Optyse LFO is considered an ideal instrument for the World Health Organization and VISION 2020 Programme, which aims to eliminate unnecessary blindness. In fact, in June 2006, the Optyse LFO was recommended for inclusion in the next issue of the VISION 2020: RIGHT TO SIGHT Standards List.

**Diabetes** Diabetes is one of the significant diseases that can be detected by examining the retina. The first sign of any diabetes in the eyes is almost always what is called 'background retinopathy'.

Background retinopathy is seen as little red spots on the retina called 'micro aneurysms', larger smudges called 'micro haemorrhages', or yellowish deposits called 'exudates'.

Background retinopathy can progress to the next stage, proliferative retinopathy, which is why monitoring by retinal exam is so valuable.

In this next stage, tiny new blood vessels form, which are fragile and liable to break and bleed. These can impair vision by bleeding into the vitreus, the eye's normal gel, which keeps the eyeball spherical, like the air does in a soccer ball.

# Lighting a whole village with one 100-watt bulb.



The Light Up The World Foundation was the first humanitarian organization to introduce solid state electric lighting to the developing world. A huge task to complete, as it's estimated that two billion people – a third of the world's population – don't have access to electricity.

Dr Dave Irvine-Halliday, who established Light Up The World in 1997, estimates that it would take 'an astronomical amount of energy' to provide conventional electric lighting to the developing world. The capital cost and the cost to the environment would be 'almost unimaginable'. An indication of the cost is that lighting accounts for 25 or 30 per cent of our total electricity consumption.

Irvine-Halliday, a Scottish-born fibre-optic expert and University of Calgary professor, was the first person to conceive that solid state lighting (SSL) could be the solution to lighting up the developing world. He and his wife, Jenny, personally financed the proof they needed in the field.

In 2001, Ken Robertson, co-founder of the LUTW Foundation, made it a personal mission to develop Irvine-Halliday's vision of a sustainable market model into a business reality.

The technology the Foundation works with is SSL created by a White Light-Emitting Diode (WLED). The WLED 'bulb' is virtually indestructible and lasts for many years. And the batteries in a WLED torch run for more than 500 hours, 10 times longer than the normal incandescent bulb torch.

Put another way, it takes the same amount of power to light a whole village as it does to run one 100-watt bulb in the developed world.

A complete LUTW lighting system is made up of three core components and installation hardware. There are two modular one-watt WLED lamps, a rechargeable, maintenance-free battery, and a five-watt solar panel.

Irvine-Halliday's mission to light up the developing world was inspired by a trekking holiday in the Annapurna Himalaya, where he was dismayed by what he saw of the villagers' living conditions. He recognized that there are

SOLAREX

SOLAR-THERM
SOLAR GENERATOR
SWITCH
POWER
LUTW LED DRIVER
www.lutw.org
700 0701 000 REV A
©2002 LUTW
ROTARY

many implications associated with light that go beyond being able to see after dark.

'I was especially concerned about the children, who often had to work all day and whose only opportunity to study was at night by kerosene light, with the attendant risks from the unhealthy fumes and the danger of fire.'

His concerns are well-founded. In 2000 alone, more than two million children died from acute respiratory disease, 60 per cent associated with indoor air pollution and other environmental factors. Apart from being unhealthy, kerosene is also expensive and environmentally damaging, too.

By contrast, LUTW's lighting brings numerous short- and long-term benefits. By allowing people to read and study after dark, Irvine-Halliday explains, 'the lighting has an enormous impact on the social, economic, physical and spiritual lives of those with little opportunity for progress. Not least of those benefits is the improvement in the education of children and women in areas where poverty and illiteracy walk hand in hand. The ability to operate a cottage industry by night using fixed or mobile task lighting helps people earn a modest living.'

LUTW, inspired by Irvine-Halliday, his wife Jenny and their son Gregor, and the dedicated office staff in Calgary, is bringing its light and many significant benefits to the deprived on every continent. So far, the Foundation has completed projects in 43 countries, helping over 100,000 people.

In January 2005, Dr Irvine-Halliday received a direct appeal from Sarvodaya, the biggest, most established NGO in Sri Lanka, to supply LUTW systems for their hastily constructed tsunami refugee camps. The initial request for 1,000 systems led to 2,000 systems being manufactured in Sri Lanka within a matter of months, and World Vision has since installed an additional 1,000 systems. The total demand could well exceed 10,000 before the project is completed.

LUTW believes that Microfinance could be highly instrumental in advancing their vision. If people from the developing world are able to buy SSL systems with Microcredit loans, a very significant step will be taken towards 'making poverty history'.



**Blackout** In August 2003, 50 million people in the north-eastern United States and south-eastern Canada got a taste of what life is like without adequate lighting.

A massive power failure threw some of the world's most sophisticated urban areas into the reality of everyday life for billions in the developing world.

As lighting, transport, air-conditioning and everything else electrical failed, millions experienced a vivid demonstration of how much the developed world takes for granted.

It was 4.11 p.m. on 14 August when the lights went out in New York. Four hours later, its famous skyline became an extraordinary silhouette against a moonlit sky. For the first time in recent history, stars could be seen over Manhattan.

## WHAT GOES ROUND COMES AROUND

I was on a speaking tour with Bob Geldof and Malcolm McLaren.

Just before we arrived in New Zealand, Brian Sweeney, a friend and major supporter of the Award sent me a TV news extract featuring Leslie Kay, who had just had an amazing breakthrough with KASPA, our very first winner.

It was great to see this amazing idea finally coming to life, and gratifying that the interview with Leslie was done with our Award statue sitting behind him.

On the tour, Bob carried in his pocket a piece entitled 'Commitment' which he would read out to the audience at the end of his presentation.

'Commitment' is by William Hutchinson Murray, from his 1951 book, 'The Scottish Himalayan Expedition'. Amazingly it wasn't the first time I'd come across it.

In the thank you letter we'd received from Dr Irvine-Halliday for our support for the Light up the World project in Sri Lanka, it was his closing piece.

It's a nice way to close this too.



## COMMITMENT

Until one is committed there is hesitancy, the chance to draw back, always ineffectiveness, concerning all acts of initiative (and creation)

There is one elementary truth, the ignorance of which kills countless ideas and splendid plans: that the moment one definitely commits oneself, then Providence moves, too.

A whole stream of events issues from the decision, raising in one's favour all manner of unforseen incidents and meetings and material assistance, which no man could have dreamed would have come his way.

Whatever you can do or dream you can, begin it.

Boldness has genius, power and magic in it.

Begin it now.

**1998 Saatchi & Saatchi Award for Innovation in Communication**

**The Ideas**

KASPA · Professor Leslie Kay SAATCHI & SAATCHI AWARD WINNER
Self-Adjustable Spectacles · Professor Joshua Silver EDWARD DE BONO MEDAL WINNER
Biochip Artificial Vision Project · Dr Mark Humayun/Dr Eugene de Juan
BioMuse · Dr Benjamin Knapp
Critical Mass Communicator · Jaron Lanier
Crossed Beam Display · Dr Elizabeth Downing
Quadkey · Dr David Levy
Quicktionary · Adi Lipman/Aharon Lipman
Seismic Detection of Tornadoes · Dr Frank Tatom
Sprint Integrated On-Demand Network (ION) · Sprint
Univers Revolved · Ji Lee

**The Judges**

**Buzz Aldrin** ***Astronaut***

Buzz Aldrin was born in Montclair, New Jersey on 20 January 1930. His mother, Marion Moon, was the daughter of an Army chaplain. His father, Edwin Eugene Aldrin, was an aviation pioneer, a student of rocket developer Robert Goddard, and an aide to the immortal General Billy Mitchell. Buzz was educated at West Point, graduating with honours in 1951, third in his class. After receiving his wings, he flew Sabre Jets in 66 combat missions in the Korean conflict, shooting down two MIG-15s. Returning to his education, he earned a Doctorate in Astronautics from the Massachusetts Institute of Technology in Manned Space Rendezvous. The techniques he devised were used on all NASA missions, including the first space docking with the Russian cosmonauts.

In October 1963, Buzz was selected by NASA as one of the early astronauts. In November 1966, he established a new record for Extra-Vehicular Activity in space on the Gemini XII orbital flight mission. Buzz has logged 4,500 hours of flying time, 290 of which were in space, including eight hours of EVA. As Backup Command Module Pilot for Apollo VIII, man's first flight around the Moon, he significantly improved operational techniques for astronautical navigation star display. Then, on 20 July 1969, Buzz and Neil Armstrong made their historic Apollo XI moon walk, thus becoming the first two humans to set foot on another world. This unprecedented heroic endeavour was witnessed by the largest worldwide television audience in history.

Upon returning from the Moon, Buzz embarked on an international goodwill tour. He was presented the Presidential Medal of Freedom, the highest honour amongst over 50 other distinguished awards and medals from the United States and numerous other countries.

Since retiring from NASA, the Air Force and his position as Commander of the Test Pilot School at Edwards Air Force Base, Dr Aldrin has remained at the forefront of efforts to ensure a continued leading role for America in manned space exploration. To advance his lifelong commitment to venturing outward in space, he has created a master plan of evolving missions for sustained exploration utilizing his concept, 'The Cycler', a spacecraft system that makes perpetual orbits between Earth and Mars. In 1993 Dr Aldrin received a US patent for a permanent space station he designed. More recently he founded his rocket design company, Starcraft Boosters, Inc., and the ShareSpace Foundation, a non-profit devoted to opening the doors to space tourism for all people.

Dr Aldrin has continued to share his vision for the future of space travel with today's youth by authoring his first illustrated children's book *Reaching for the Moon* (HarperCollins, 2005) that debuted at No. 2 on the *New York Times* Bestseller's List, and has garnered many publishing accolades including the *Parent's Choice* Recommended Award and *Publisher's Weekly* Best Children's Book of 2005. He has authored two space novels that dramatically portray man's discovery of the ultimate frontier: *The Return* (Forge Books, 2000) and *Encounter with Tiber* (Warner Books, 1996). He has also authored an autobiography, *Return to Earth* (Random House, 1973) and the bestseller historical documentary, *Men from Earth* (Bantam, 1989), describing his trip to the Moon and his unique perspective on America's space programme.

On Valentine's Day 1988, Buzz married Lois Driggs of Phoenix, Arizona. She is a Stanford University graduate, an active community leader in Southern California, and personal manager of Buzz's commercial enterprises. Their combined family is comprised of six grown children and one grandson. Their leisure time is spent exploring the deep-sea world of scuba diving and skiing the mountaintops of Sun Valley, Idaho.

Now Buzz, as Starcraft Enterprises – the name of his private space endeavours – lectures and travels throughout the world to pursue and share his latest concepts and ideas for exploring the Universe. He is a leading voice in charting the course of future space efforts from Planet Earth.

**Laurie Anderson** ***Multi-Media Artist***

'What is Laurie Anderson?' is not an entirely unreasonable question.

Her artistic career has been a swervy journey from visual artist to composer to poet to photographer, to ventriloquist, to electronics whiz to filmmaker to vocalist and instrumentalist.

It started with the violin, which she played in her hometown's Chicago Youth Orchestra.

She moved on to graduate in Art History from Barnard College in New York in 1969, and then obtained an MFA in sculpture from Columbia University.

Downtown New York in the early 70s was a place full of artistic experimentation, to which Laurie was drawn.

Her earliest performances tended to be on the street or in informal art spaces. Perhaps the most notable of her performances at this time involved her standing on a block of ice wearing ice skates, and playing her violin. The performance ended when the ice melted.

Since this bi-media work with ice and violin, Laurie has gone on to multi-media creations of complexity, theatricality, surprise and originality. Harnessing music, video, storytelling, projected imagery and sculpture, Laurie emerges as an electrifying performer.

In 1982 her single 'O Superman' brought Laurie to the attention of a far wider audience. It reached No. 2 in the British pop charts, after the eclecticism-driven British DJ, the late John Peel, started playing the record.

The single came from her debut album *Big Science* which has been followed by six further albums.

Her profile as a visual artist is impressive too, having been exhibited at the Guggenheim Museum in Soho, New York, as well as extensively in Europe, including the Centre Georges Pompidou in Paris.

Her retrospective show, *Sound in the Work of Laurie Anderson*, has been appearing in European galleries and museums for the last three years.

In 1999 she staged *Songs and Stories From Moby Dick*, an interpretation of Herman Melville's novel.

And in 2005 she toured North America, Japan and much of Europe with her performance-based piece, *The End of the Moon*.

Which leads very naturally to mention another of Laurie's pioneering roles, that of NASA's first artist-in-residence.

The list of her collaborators is extraordinary. This includes, William Burroughs, Mitchell Froom, Peter Gabriel, Perry Hoberman, David Sylvian, Jean Michel Jarre, Nona Hendryx and Lou Reed, her long-term romantic partner.

Something of a Who's Who answer to the 'What is Laurie Anderson?' question.

**Edward de Bono** ***Lateral Thinker***

Dr Edward de Bono is one of the most celebrated residents in the world of ideas.

He invented the concept of 'lateral thinking' and the term is now an entry in the Oxford English Dictionary correctly attributed to him.

He's regarded as a leading international authority on conceptual and creative thinking, innovation and the direct teaching of thinking as a skill. Edward advises various governments, cities, regional governments and global organizations. He deals on a macro level with diverse topics, including economy, unemployment, social policy, recidivism, pensions, healthcare, finance, transportation, education, conflict resolution, judicial processes and foresight scenario design.

Edward was born on the island of Malta in 1933 and has now established his World Centre for New Thinking there.

The Centre acts as a platform and channel to make visible new thinking from any source. Edward believes very strongly in the power of creative thinking to solve problems and conflict. But as Edward observes, 'democracies and representative organizations, due to their nature, cannot put forward new ideas. By definition, "new ideas" are not representative of existing thinking. They are therefore high risk. Such organizations may be perfectly capable of having new ideas but cannot risk putting them forward.'

Edward's fundamental belief is that anyone can be a creative thinker, and that it's simply a matter of learning and developing their skills.

He has created pioneering systems to encourage creative thinking, the most famous being Lateral Thinking and the Six Thinking Hats technique.

Many corporations and institutions benefit from employing these methods on a constant basis.

Edward is an enemy of complexity and a dedicated advocate of simplicity. In fact, he's dedicated a whole book to the subject: *Simplicity*.

He's written 69 other books as well, with translations into 40 languages. These include *Lateral Thinking, Serious Creativity, The Mechanism of Mind, Handbook for a Positive Revolution* and *The de Bono Code Book*.

His methods for 'thinking' are mandatory on the school curricula of many countries and are widely used in others. Countries include Australia, New Zealand, Canada, Argentina, UK, Italy, UAE, Ireland, Spain, Portugal, the Baltic States, Sweden, Denmark, Norway, Singapore, Malaysia, India, China, USA, Russia and Venezuela.

As regards his own education, Edward was a Rhodes Scholar at Oxford. He holds an MA in Psychology and Physiology from Oxford, a D Phil in Medicine and a PhD from Cambridge. He has also been awarded a D Des (Doctor of Design) from the Royal Melbourne Institute of Technology, and an LLD from Dundee. He holds professorships at the universities of Malta, Pretoria, Dublin City and Central England. The New University of Advancing Technology in Phoenix, Arizona appointed Dr de Bono their Da Vinci Professor of Thinking in 2005.

Edward is also King of Geraldton, Margaret River in Australia, and of Launceston in Tasmania. Why? Because Edward thought it was a good idea, so he wrote a book *Why I should be King of Australia*, and the people of Geraldton and Launceston obviously agreed.

**James Burke** ***Author/TV Presenter***

James Burke is a master communicator. He can make science palatable, information absorbable and complex theories not only understandable but enjoyable too.

Burke is perhaps best known for his 1979 TV series *Connections*, a science-history 10-parter, filmed in over 150 locations in more than 19 countries.

The series explored seemingly unrelated events, situations and people, and fitted them into a puzzle that helped explain the fundamental process of social and technological change.

As it happens, the theory of *Connections* could be applied to James's story.

He was born in Northern Ireland in 1936. He gained an MA in English from Jesus College, Oxford. He moved to Italy where he lectured at Bologna and Urbino Universities and at various English schools.

He was involved in creating an English–Italian dictionary and in publishing an art encyclopedia.

In 1966, he moved to London and BBC Television to co-host *Tomorrow's World*, a weekly prime-time science magazine programme.

His one-man award-winning TV science show, *The Burke Special*, ran from 1972 to 1976.

Which brings us right back to *Connections*. The programme achieved the highest-ever US audience for a documentary series, and is on the curriculum of some 350 US colleges and universities. From technology, James turned his attention to the brain and human perception for *The Real Thing*. This six-part series was seen in 30 countries.

During the 1980s, James produced and hosted more landmark programmes and events. These included the 10-part *The Day The Universe Changed*, the coverage of the return of Halley's Comet, and the provocative two-hour mini-series, *After The Warming*.

Another lateral shift saw James, in 1991, host *Masters of Illusion*, a programme about Renaissance painting.

Between 1991 and 1999, James returned to technology and produced the 20-part *Connections 2*, and the 10-part *Connections 3*.

James's bestselling books include *Twin Tracks, The Pinball Effect* and *The Axemaker's Gift* and, of course, *Connections. American Connections*, about the signatories of the US Declaration of Independence, is due to be published in 2007.

James is the founder of the Burke Institute for Innovation in Education.

He has received several honorary doctorates for his work, and *The Washington Post* has described him as having 'one of the most intriguing minds in the Western World'.

**William Gibson** ***Sci-fi Novelist***

William Gibson has enjoyed that rare pleasure for a fiction writer. He made something up and has watched it gradually become reality.

His first novel, *Neuromancer*, published in 1984 (appropriate) is the original cyberpunk novel.

In it, Gibson presented the idea of a global information network (called the Matrix) and gave the world the term 'cyberspace'.

William was born in 1948 on the coast of South Carolina. His father died when he was six. His mother when he was 18.

But William is a writer. His own words tell his story better than we ever could …

'It was a world of early television, a new Oldsmobile with crazy rocket-ship styling, toys with science-fiction themes. Then my father went off on one more business trip. He never came back. He choked on something in a restaurant, the Heimlich Maneuver hadn't been discovered yet, and everything changed.

'My mother took me back to the small town in southwestern Virginia where both she and my father were from, a place where modernity had arrived to some extent but was deeply distrusted. The trauma of my father's death aside, I'm convinced that it was this experience of feeling abruptly exiled, to what seemed like the past, that began my relationship with science fiction.

'At age fifteen, my chronically anxious and depressive mother having demonstrated an uncharacteristic burst of common sense in what today we call parenting, I was shipped off to a private boys' school in Arizona.

'In Arizona, science fiction was put aside with other childish things, as I set about negotiating puberty and trying on alternate personae with all the urgency and clumsiness that came with that, and was actually getting somewhere, I think, when my mother died with stunning suddenness. Dropped literally dead: the descent of an Other Shoe I'd been anticipating since age six.

'Thereafter, probably needless to say, things didn't seem to go very well for quite a while. I left my school without graduating, joined up with the rest of the Children's Crusade of the day, and shortly found myself in Canada, a country I knew almost nothing about. I concentrated on evading the draft and staying alive, while trying to make sure I looked like I was at least enjoying the Summer of Love. I did literally evade the draft, as they never bothered drafting me, and have lived here in Canada, more or less, ever since.

'Having ridden out the crest of the Sixties in Toronto, aside from a brief, riot-torn spell in the District of Columbia, I meet a girl from Vancouver, went off to Europe with her (concentrating on countries with fascist regimes and highly favorable rates of exchange), got married, and moved to British Columbia, where I watched the hot fat of the Sixties congeal as I earned a desultory bachelor's degree in English at UBC.

'In 1977, facing first-time parenthood and an absolute lack of enthusiasm for anything like "career," I found myself dusting off my twelve-year-old's interest in science fiction. Simultaneously, weird noises were being heard from New York and London. I took Punk to be the detonation of some slow-fused projectile buried deep in society's flank a decade earlier, and I took it to be, somehow, a sign. And I began, then, to write.

'And have been, ever since.

'I suspect I have spent just about exactly as much time actually writing as the average person my age has spent watching television, and that, as much as anything, may be the real secret here.'

**Tibor Kalman** ***Designer***

'I use contrary-ism in every part of my life. In design … I'm always trying to turn things upside down and see if they look any better.'

This was Tibor Kalman talking in December 1998, five months before he died.

His 30-year career is a mosaic representation of his restless curiosity and subversive wit.

In 1979 he founded the multi-disciplinary design firm M&Co. The company created everything from watches to album covers, notably for Talking Heads.

In 1991, Tibor closed M&Co's New York offices to accept Oliviero Toscani's invitation to create a magazine.

Toscani (see page 253) was the creative director of Benetton, the Italian fashion brand, and his provocative advertising images had generated acres of publicity for the company.

Toscani wanted Tibor to collaborate in the creation of a magazine that embodied Benetton's radical chic ethos, without being an overt mouthpiece.

What emerged was *Colors*.

After working on the magazine in New York for a few years as editor and designer, Tibor moved, with his wife Maira and his two children, to Rome, to continue producing.

The magazine was hugely innovative, inspiring and influential. It relied heavily on graphics and a minimum of text. It's said that *Colors* was the perfect platform for Tibor's visual and philosophical ideas, and he was responsible for the first 13 issues.

His passion for breaking new ground in design was matched by a passionate commitment to social causes.

He believed that members of the American Institute of Graphic Artists should be required to do charitable work.

He left New York University for a while, when he was an undergraduate, to support the Communists in Cuba.

This may be a little ironic since Tibor and his parents were forced to leave Budapest, where he was born in 1949, in order to escape the 1956 Soviet invasion.

The relationship with Toscani fell apart. Tibor left *Colors*, returned to New York and reopened M&Co.

The first symptoms of the cancer that killed him had already been detected.

The book *Perverse Optimist* started as a retrospective of Tibor's work. It became his legacy.

Matthew Haber: 'When designer Tibor Kalman died of non-Hodgkin's lymphoma on May 2nd [1999] in Puerto Rico, surrounded by his wife, Maira, and family, he died as he had lived and worked – on his own terms and with the generosity of spirit and optimism that touched everyone who knew him.'

**Lachlan Murdoch** ***Media Executive***

On 29 July 2005, Lachlan Murdoch sprang a major surprise on the media world by resigning his executive roles with News Corporation.

Lachlan is the elder son of media mogul Rupert Murdoch, and was widely seen, until 29 July 2005, as the heir apparent to Rupert.

Lachlan had reached this position by learning about the business from the inside. And from the bottom. His first experience was as a printing press cleaner on weekends when he was in high school in the US.

He earned a BA in Philosophy from Princeton and moved to Australia in 1994 to begin his training in the family's Queensland Newspapers.

His image at the time didn't quite fit the corporate expectation. He was a rock climber with a lizard tattoo, riding his pricey motorcycle around Sydney. He soon recognized the need to modify his image. 'I have to prove I'm serious,' he said in 1995. Two years later he was promoted to Chief Executive of News Limited.

His next promotion, in 1999, brought Lachlan back to the US to run News Corporation's print operations, including the *New York Post* and the HarperCollins Publishing Group.

He was now in the inner sanctum of News Corporation.

A further promotion in October 2000 took Lachlan another step towards succession when he became News Corporation's Deputy Chief Operating Officer.

People began to compare Lachlan favourably with Rupert.

Fox Television's stations chairman, Mitch Stern, said Lachlan is 'a lot like him ... he's engaging and curious. Intelligent. And, I'd say – and this is where they stand out from the rest of the industry – a gentleman with a high sense of integrity. Pretty good drinker, good taste in wine.' Early in 2004 Lachlan added another responsibility, taking over the Chairmanship of Fox TV Network.

Then in 2005, aged 33, Lachlan chose to move to Australia for his wife, Sarah O'Hare and one-year-old son Kalan. Within weeks he was setting up Illyria Pty Ltd, in which he acts as Sole Director and Secretary.

**Richard Saul Wurman** ***Information Architect***

Richard Saul Wurman was 26 years old when his first book was published in 1962. It launched the singular passion of his life: Wurman wants to make information understandable.

Each of his subsequent 81 books focuses on a topic or idea that he himself had difficulty understanding.

Interestingly, therefore, they're based not on what he already knows but on his desire to know, on his ignorance rather than his intelligence and on an inability rather than an ability.

Being an architect and a graphic designer, Richard writes. He also designs the books.

His highly popular ACCESS travel guides exemplify his thinking about information.

The guides organize content by neighbourhood. This innovative approach is based on how we look for information, which in this case, is by location.

So, for example, instead of grouping all the hotels together and all the shops together in different chapters, a shop next to a hotel is how information about both appears in the guide. The simple but effective use of coloured text makes it easy for readers to separate, locate and evaluate restaurants, museums, parks and other categories of places of interest for tourists.

Richard's emphatically modest assertion that 'My expertise is my ignorance', and his refusal to credit himself with having any ideas, both have a hollow ring to them.

His 1962 recognition about the need to manage information effectively was extremely prescient. Today, information proliferates at an unprecedented pace. But Richard was already on to it in 1976 when he coined the term 'information architect' when he observed the massive amount of information society was generating with little care or order.

On the subject of having ideas, Richard must have forgotten that he created the TED Conferences in 1984, now described as 'the ultimate brain spa'.

His best-selling books include *Information Anxiety* and *Information Anxiety 2* and the *Understanding* series: *Understanding USA*, *Understanding Healthcare* and *Understanding Children*.

**2000 Saatchi & Saatchi Award for Innovation in Communication**

**The Ideas**

QTC (Quantum Tunnelling Composites) · David Lussey SAATCHI & SAATCHI AWARD WINNER
DigiScents · Joel Bellenson/Dexster Smith EDWARD DE BONO MEDAL WINNER
Digital Building Blocks · Gisue Hariri/Mojgan Hariri
Biomechanical Transducer (Electric Shoe) · Trevor Baylis/Barry James/John Monteith
LSTAT (Life Support for Trauma & Transport) · Integrated Medical Systems Inc.
Rosetta Disk · The Long Now Foundation
Universal Pill Marking System · Dr Gerhard Ovellhaus/Richard Peterson
Wireless Capsule Endoscope (Pillcam) · Gavriel Iddan/Gavriel Meron/Tim Mills/Professor Paul Swain

**The Judges**

**Edward de Bono** ***Lateral Thinker*** *(page 237)*

**Paul Davies** ***Theoretical Physicist***

British-born Professor Paul Davies is a theoretical physicist, cosmologist, astrobiologist, author and broadcaster. He is College Professor at Arizona State University in Phoenix. Previously he was Professor of Natural Philosophy in the Australian Centre for Astrobiology at Macquarie University. He has also held academic positions at the universities of Cambridge, London, Newcastle Upon Tyne and Adelaide.

His research has ranged from the origin of the Universe to the origin of life, and includes the properties of black holes, the nature of time and quantum field theory. He's the writer of over 25 popular and academic books. Titles include *How To Build A Time Machine, The Mind of God,* and *The Last Three Minutes.*

Given his field, it's probably impossible for Paul to not think big. In the early 1990s for example, he became interested in the controversial theory that comets hitting planets could blast material into space, opening up the possibility that rocks might travel from one planet to another.

Paul says, 'At the same time, I learned about an equally controversial theory, that life on Earth extends deep into the crust, in the form of microbes dwelling kilometres underground inside apparently solid rock. Putting two and two together, it dawned on me that rocks flung off Earth by impacts could convey microbes living in them to Mars, and vice versa. Though it took many years for this scenario to become accepted, it is now generally acknowledged as a distinct possibility.'

This theory formed the central theme of another Davies book, *The Origin of Life.*

As a broadcaster, Paul has created some highly regarded programmes for radio and television.

One of the series of BBC Radio 3 science documentaries called *Desperately Seeking Superstrings* won the Glaxo Science Writers Fellowship when it was transcribed into a book.

In 2000 he devised and presented a three-part BBC Radio 4 series on the origin of life, *The Genesis Factor.*

His TV projects include two six-part Australian series, *The Big Questions* and *More Big Questions*, and a BBC 4 documentary in 2003 about his astrobiology work, *The Cradle of Life.*

He has received a number of awards for his media work. These include the 2001 Kelvin Medal from the UK Institute of Physics and the 2002 Michael Faraday Prize from the Royal Society for his contribution to promoting science to the public.

He also won the 1995 Templeton Prize, the world's largest annual prize, for his work on science and religion.

Paul has always been fascinated by the idea that we may not be the only intelligent life in the Universe. He also wonders whether life has emerged more than once on Earth and that these 'aliens' are still with us.

He's also an advocate for establishing a permanent human presence on Mars.

His focus now is astrobiology, a new field of research seeking to understand the origin and evolution of life, and to search for life beyond Earth.

In this context it seems entirely fitting that, in April 1999, asteroid 1992 OG was officially named (6870) 'Pauldavies' in his honour.

**Brian Eno** ***Musician***

Brian Eno has nearly as many names as he has interests.

Brian Peter George St John Le Baptiste de la Salle Eno is a singer, songwriter, futurologist, professor, curator, artist, theorist, diarist, political activist, fundraiser, genre inventor, multi-media artist and record producer.

He was born in 1948 in Suffolk, England, and for decades called himself a non-musician. Indeed, in 1968 he wrote the limited edition theoretical handbook, *Music for Non Musicians.*

After studying at art schools in Ipswich and Winchester, Brian moved to London, lived in an art commune and played in, amongst others, the Portsmouth Sinfonia. The Portsmouth was a glorious cacophony created by an orchestra of players who apparently could only just play their instruments. Brian played clarinet and produced both their albums.

He joined the British glam-rock band Roxy Music as 'technical adviser' in January 1971. Before he left in 1973 he'd imbued the band's work with some startlingly original electronic touches.

His solo career got off to a successful start. His album *Here Come the Warm Jets* was a particular highlight.

His concept of ambient music came about by accident. Literally.

Brian was in hospital after being knocked down by a taxi in London. A friend, Judy Nylon, visited him and put on a record of harp music. When Judy left, Brian couldn't get out of bed to turn up the volume when a rainstorm almost drowned out the sound of the music, which was coming from the one working speaker.

After initial annoyance, Brian decided to create music 'in a way that you might use light or colour, or a painting on the wall'. The albums *Another Green World* and *Discreet Music* followed.

Ever the dilettante (a badge he wears with pride, incidentally), Brian collaborated with painter Peter Schmidt on *Oblique Strategies*, a series of cards designed to encourage lateral thinking.

Brian's output of ambient music continued between 1978 and 1990, with *Music for Airports* perhaps the most influential.

Collaborations with David Bowie on *Low, Heroes* and *Lodger* and with John Cale, the former Velvet Underground member, on *Wrong Way Up*, dovetailed with Brian's increasing reputation and output as a producer. Brian has been described as 'the liquid engineering in the manufacture of contemporary music'.

The list of artists he has produced or appeared with is incredible. These include U2, John Cale, David Bowie, Laurie Anderson, Johnny Cash, Elvis Costello, Depeche Mode, Peter Gabriel, Talking Heads and Robert Wyatt.

The highly influential, *My Life In The Bush Of Ghosts* which he created and produced with David Byrne (see page 246) was reissued in 2006, with additional previously unheard tracks, to mark its 25th anniversary. The reissue was groundbreaking, just as the original had been. A dedicated website offered visitors downloadable access to all the multi tracks on two of the songs, allowing anyone to remix and sample the material.

Brian's latest solo album, *Another Day On Earth* was released in 2005. His credit on Paul Simon's 2006 release, *Surprise*, is for 'Sonic Landscape'.

Brian's influence on the artistic avant-garde has been further enhanced by his recent *77 Million Paintings*. This is a generative televisual installation that slowly reconfigures a collection of images through its own software engine. Played at its quickest generative pace, it would take 9,000 years to watch in its entirety. Played at its slowest, it would take several million years.

'Slow' is not new to Brian. He co-founded The Long Now Foundation, which encourages people to think in the very long term.

**Kevin Kelly** ***Writer***

*Wired* magazine describes itself as the 'voice of the digital revolution'.

Kevin Kelly helped launch it in 1993, and until January 1999 was its Executive Editor. He's now *Wired*'s Editor-At-Large.

At one point, the *New York Times* described Kevin as *Wired*'s 'Big Think Guy', bringing depth and fresh ideas to something that might otherwise be dominated by 'attitude'. The fact that Kevin doesn't like computers could help in this.

But there's a lot more to Kevin than *Wired*. He used to be a nomadic photojournalist. One summer he cycled right across America. And for most of the 70s he was taking photographs in remote parts of Asia and having them published in national magazines.

He wrote a monthly travel column for *New Age Journal*, and in the early 1980s he published the first magazine dedicated to walking.

By his own admission he's an incurable magazine junkie.

In the late 1980s he created four versions of *The Whole Earth Catalogs*. These award-winners evaluate all the best 'tools' available for self-education.

Kevin is a member of the Global Business Network, a consulting group specializing in creating scenarios of the future for global businesses.

He is a Fellow at the Center for Business Innovation, run by Ernst & Young.

He's passionate about a campaign to make a full inventory of all living species on Earth. Most taxonomic groups endorse the All Species Inventory as an idea whose time has come.

The aim is to make a web-based complete catalogue in one generation. In other words, in the next 25 years.

Kevin is a member of the board of The Long Now Foundation (a name coined by Brian Eno). The Foundation is a non-profit organization that encourages 'slower, better' thinking in the hope of fostering long-term responsibility (see Rosetta Disk, page 77).

Kevin, through the Foundation, has made a number of predictions. These include the suggestion that by 2085, China will be considered a Christian nation with at least 33 per cent of its population calling themselves Christians.

Kevin's take on the future led the makers of the film *The Matrix*, to require their principal actors to read three books before filming started. One was Kevin's *Out of Control: The New Biology of Machines, Social Systems, and the Economic World.*

When it was first published, *Fortune* magazine called it 'essential reading for all executives'.

Kevin Kelly was born in 1952, and lives in a small coastal town, Pacifica, just south of San Francisco.

**Pattie Maes** ***Artificial Intelligence Professor***

As a child in Belgium, Pattie Maes wanted to be a vet.

Instead she's an associate professor in MIT's Program in Media Arts and Sciences.

She founded and directs the Media Labs Ambient Intelligence research group.

Before she joined the Media Lab, Pattie was a visiting professor and a research scientist at the MIT Artificial Intelligence Lab.

She founded and ran the Software Agents group and was something of a pioneer in what was then a relatively new research area.

Software Agents are semi-intelligent computer programmes that assist a user with the overload of information and the complexity of the online world.

The connection between this and her original vetinerary ambition may not be entirely obvious. But there is one. Pattie says she involved herself in Software Agents because of her own information overload. And this started with animals.

'Chickens and rabbits and fish,' according to her mother. 'She bred rabbits. She always brought home sick cats and dogs and so on ... she wanted to be a vet ... it was only later she became interested in computers.'

She holds bachelor's and PhD degrees in Computer Science from the Vrije Universiteit Brussel in Belgium.

Her areas of expertise are human-computer interaction, artificial life, artificial intelligence, collective intelligence, and intelligence augmentation.

She found herself invited to MIT initially by Marvin Minsky, one of the pioneers of artificial intelligence, because of her passion for the subject.

A two-month visit turned into a visiting professorship, which in turn, etc.

Pattie is the editor of three books. She's an editorial board member and reviewer for numerous professional journals and conferences.

*Newsweek* identified her as one of the '100 Americans to watch for' in 2000. *Time* magazine placed her as a member of the Cyber-Elite, the top 50 technological pioneers of the high-tech world. The World Economic Forum honoured her with the title 'Global Leader For Tomorrow'.

Pattie says she went into computer science because technology factored into so many disciplines, such as business, architecture, science and art.

She's deeply interested in these fields. She spends some of what free time she has going to art galleries. 'I enjoy sculpting and photography and tennis and swimming. I simply do not have enough time. My life is not in control.' Which is probably more of an observation than a complaint.

**Kjell A. Nordström** ***Economist***

With shaved head, all-black wardrobe and small horn-rimmed spectacles, Swede Kjell Nordström looks more like an architect than an economist.

Read *Funky Business*, or the more recent *Karaoke Capitalism*, both co-written with his co-guru Jonas Ridderstråle (shaved head, all-black wardrobe, small horn-rimmed spectacles) and he doesn't sound like an economist either.

The style is punky and punchy. Littered with snappy new terms. Short sentences. Instant and get-it. Kjell, whose seminars are sell-outs and described as 'gigs', is a guru who's an academic.

He holds an Economic Licentiate degree and a doctoral degree from the Stockholm School of Economics.

He is an Associate Professor at the Institute of International Business (IIB) at the Stockholm School of Economics.

The 2005 Thinkers 50, the world's first ranking of management thinkers, placed Kjell and Jonas at No. 1 in Europe and No. 9 in the world.

Kjell's ideas revolve around globalization and innovation.

He identified the sea change that has shifted competitive advantage from what an organization makes to the way its people are able to think.

Brain has overtaken brawn.

The one and a half kilograms of grey matter between our ears outweigh the tonnes and tonnes of metallic matter sitting in factories around the world.

Being different is the key to business survival.

The book *Karaoke Capitalism: Management for Mankind* explores this theme further.

The Karaoke economy is dominated by people with endless choices. Unfortunately for business, the Karaoke Club is also home to institutionalized imitation. Imagination and innovation are vital for survival, let alone prosperity.

When Kjell takes his thinking on the road with Jonas the impact is profound.

Again, the content belies the dry economics tag.

According to the *London Times*, Kjell and Jonas 'bring a slice of rock 'n' roll rebellion to the international speaking circuits'.

When Kjell isn't on the road, he's responsible for the International Business Course at the Stockholm School of Economics.

He's one of the founders of the 'Advanced Management Program'. AMP is the School's most prestigious offering and it attracts Scandinavia's top management talent.

Unsurprisingly, Kjell is also on the board of directors of several companies.

*Funky Business* has been ranked as the sixteenth best business book of all time.

The fact that it's been translated into more than 31 languages underlines the universality of Kjell's economic wisdom.

**2002 Saatchi & Saatchi Award for Innovation in Communication**

**The Ideas**

Light Up The World Foundation · Dr Dave Irvine-Halliday SAATCHI & SAATCHI AWARD WINNER
FanWing · Patrick Peebles EDWARD DE BONO MEDAL WINNER
Artificial Vision for the Blind · Dr William Dobelle
Global Earthquake Monitoring System (GEMS) · Professor Stuart Crampin
Intelligent Mobile Phone Charger · Paul Stokes
Kinkajou Projector · Design that Matters
Mind Switch · Dr Ashley Craig
NeuroGraph Diagnostic Aid System · Dr Richard Granger
SmartSlab · Professor Tom Barker
Solar-Powered Mobile Phone Charger · Richard O'Connor
Stand Up And Walk Project · Professor Pierre Rabischong

**The Judges**

**Edward de Bono** ***Lateral Thinker*** *(page 237)*

**David Byrne** ***Multi-Media Artist***

David Byrne is an artistic pioneer.

And along the way, he's collaborated with many fellow pioneers. Brian Eno, Twyla Tharp, Robert Wilson and Bernardo Bertolucci included.

The globally successful rock band he led, Talking Heads, sounded like no one else with their wry, observational, no-barrier lyrics coupled to rocky, yet odd melodies.

He's a champion of world music, setting up his own world-music record label, Luaka Bop, in 1988, with a stable of brilliant talent including Cornershop, Os Mutantes, Los De Abajo, Jim White, Zap Mama and Tom Zé.

He can be contrary too. He chose to give an article he wrote for the *New York Times* the title 'I Hate World Music'.

In 1986, he directed his first feature film, *True Stories*. And he won an Oscar for Best Original Score when he collaborated with Ryvichi Sakamoto and Cong Su to create the soundtrack for Bernardo Bertolucci's *The Last Emperor* in 1987.

Working with choreographer Twyla Tharp in 1981, they created *The Catherine Wheel*, a ballet featuring unusual rhythms and lyrics.

David is a highly successful photographer with exhibitions mounted around the world since the 1990s. He enslaved technology for artistic expression when he released *Envisioning Emotional Epistemological Information* in 2003. This is a book and DVD of artwork composed entirely in Microsoft PowerPoint.

His book of tree drawings, *Arboretum*, was published in 2006.

With musical contributions from Fatboy Slim, David's current project, about Imelda Marcos, is *Here Lies Love*.

*My Life In The Bush Of Ghosts*, David's groundbreaking collaboration with Brian Eno (see page 242), was re-released in 2006 to mark the 25th anniversary of the original. The album is rightly recognized as a significant step forward in the use of sampling as a legitimate musical form.

The re-released version included previously unheard tracks. And it broke new ground again, by offering, through a dedicated website, total downloadable access to all the multi tracks on two of the songs, enabling anyone to remix and sample the material.

David Byrne was born in Dumbarton, Scotland in 1952, moved with his family to North America when he was two, and currently lives in New York.

**W. Daniel Hillis** ***Inventor***

The artificial intelligence pioneer Marvin Minsky says, 'Danny Hillis is one of the most inventive people I've ever met, and one of the deepest thinkers'. He also says, 'I think he's now on the road to becoming one of our major evolutionary theorists'.

The fact that Danny holds over 50 US patents certainly justifies the inventive tag.

Amongst other things, he pioneered the concept of parallel computers, the basis for most super-computers.

He was motivated to explore this area when he was a student at MIT and was examining the limitations of computation and the possibility of building highly parallel computers.

The result was his 1985 design of a parallel computer with 64,000 processors. He named it the Connection Machine, and it became the topic of his PhD.

He co-founded Thinking Machines Corp to produce and market the Connection Machine. Customers included American Express and NASA.

Hillis left Thinking Machines in 1995 to form DHSH, a small consulting company.

One of DHSH's clients was the Walt Disney Company, and in 1996 Hillis joined Disney full time in the newly created role of Disney Fellow.

Danny stepped into the role saying, 'I've wanted to work at Disney ever since I was a child'. He added, 'Now I have the perfect job – bringing computer magic into Disney'.

He applied his magic, designing new theme-park rides and a full-sized walking robot dinosaur. Although this heady mixture of brilliant science and massive fun must have been exhilarating and distracting, Danny didn't stop thinking big and universally.

His interest in the far-distant future led Danny to jointly set up The Long Now Foundation (see page 77) after designing the Clock of the Long Now.

He described the clock in manifesto style in 1993: 'I would like to propose a large (think Stonehenge) mechanical clock, powered by seasonal temperature changes. It ticks once a year, bongs once a century, and the cuckoo comes out every millennium.'

When Hillis left Disney he started up Applied Minds, Inc. and became its Chairman and Chief Technology Officer.

Applied Minds is a research and development company creating a range of new products and services in software, entertainment, electronics, biotechnology and mechanical design. The company also provides advanced technology, creative design and consultancy services to a variety of clients.

Danny Hillis is a Fellow of the American Academy of Arts and Sciences, a Fellow of the Association of Computing Machinery, a Fellow in the International Leadership Forum, and a member of the National Academy of Engineering. He also advises the US government.

He's received the Hopper Award for his contributions to computer science, the Ramanujian Award for his work in applied mathematics, the spirit of American Creativity Award for his inventions, and the inaugural Dan David Prize for shaping and enriching society and public life.

He wrote the book *The Pattern On The Stone: The Simple Ideas That Make Computers Work.* It's been described as 'a welcome book that makes a computer seem as basic as a bicycle'. Which could only be achieved by someone with such a truly profound understanding of computers.

**Kenji Kitatani** ***Media Executive***

Dr Kenji Kitatani is considered an expert in the US and Japanese entertainment and media business landscapes, and his 1991 book, *American CATV*, is regarded as essential reading for media professionals in Japan.

He achieved this status through a wide variety of roles in both areas. For example, back in 1981–82, he wrote 52 English scripts for the animation series *The Astro Boy*.

During the 1980 US presidential election he was special correspondent for TV Asahi Network, and produced several PBS documentaries.

From 1991 to 2003, Dr Kitatani was a member of the Board of Directors of Tokyo Dome Corporation, Japan's leading leisure and entertainment conglomerate.

He was also President of Tokyo Dome Enterprises Corporation. At the Dome, Dr Kitatani was responsible for putting on an impressive range of sports events, including the annual NFL pre-season series of games, and concerts featuring super talents such as the Rolling Stones, Michael Jackson, David Bowie and Madonna. Through his negotiations with the PGA Tour, Tokyo Dome Corporation became the sole licensor of official TPC golf courses in Japan.

He also changed the entertainment industry forever.

He believed that promoters selling tickets for events were missing an opportunity.

Dr Kitatani came up with the idea of the complete hospitality package, to include merchandise, a programme, backstage passes and a meet-the-artist opportunity.

'The artists' management saw the value of my idea immediately, but typical ticket agencies didn't get it,' Dr Kitatani says. 'They were so fixed on selling tickets, and their infrastructure wasn't made for anything more.'

But resistance soon collapsed, and Dr Kitatani's 'Golden Circle' tickets soon became the industry standard.

Advances in technology at the time helped, since much of the contact with 'Golden Circle' customers is through mobile phones and the Internet.

Dr Kitatani also worked for the Tokyo Broadcasting System (TBS) as its counsel on International Affairs and was President of its Media Research Institute for 15 years until 1999. He negotiated contracts and coordinated coverage for many major sports events including the SuperBowl and the Indianapolis 500.

From 1987 to 1994 he also managed the Hollywood film and Broadway show investment funds for TBS.

In October 1999, Dr Kitatani joined Sony Corporation as Executive Strategist, Media Content, Broadcasting and Communications. In May 2001, he was appointed Executive Vice President, Business Planning, Sony Corporation of America.

In 2004, Dr Kitatani returned to Japan to be near his ailing parents, but he remained with Sony as an Executive Advisor until 31 March 2006, when he retired from the company.

Dr Kitatani is currently Lester Smith Distinguished Professor of Media Management, The Edward Murrow School of Communication, Washington State University. He is also on the boards of PIA Corporation, the largest entertainment and sports ticketing service company in Japan, and of Optigenex, a New York-based public company with anti-ageing DNA repair technology patents.

In addition, Dr Kitatani was the executive producer of 'Terje 2006', a large media event staged in Yokohama. The event was sponsored by the Kingdom of Norway to commemorate the 100th year of Henrick Ibsen's passing.

**John Maeda** ***Graphic Designer/Computer Scientist***

John Maeda must have a perfectly balanced brain, with his left side and right side contributing equally. John, after all, is a graphic designer and artist and a computer scientist.

He's world-renowned for his work at this art-science crossroads, and was named by *Esquire* magazine as one of the 21 most important people for the twenty-first century.

John is at the MIT Media Lab, where he's Associate Professor of Design and Computation, the E. Rudge and Nancy Allen Professor of Media Arts and Sciences, and a co-Director of the Lab's Physical Language Workshop and of its SIMPLICITY Consortium.

SIMPLICITY's work is 'a radical re-examination of ways to break free from the intimidating complexity of today's technology and the frustration of information overload'.

SIMPLICITY has been a constant theme through John's career. He's been described as a founding voice for its application in the digital age. He's currently working on 'Ten Laws of Simplicity'. He started with 16, but in the true spirit of the project he's refining and reducing. Ultimately, to have just One Law would be ideal.

John received both his BS and MS degrees from MIT and he earned his PhD in design from Tsukuba University Institute of Art and Design in Japan. He's currently working on his MBA in his spare time.

John's early work redefined the use of electronic media as a tool for artistic expression, using the computer as an artistic medium in its own right, rather than a substitute for brush and paint.

John is credited, through this work, with helping to pioneer the interactive motion graphics we now see being used so widely on the Internet.

His book *Design By Numbers* captures this marriage of art and science in its title. John describes the book as 'a simple language, designed to introduce the most basic of ideas of computer programming in the context of drawing'.

Design By Numbers is also the name of a global project, initiated by John, to teach computer programming to visual artists through a freely available, custom software system that he designed.

John continues to be a pioneer. He's currently leading an initiative to 'redesign technology' so that it 'consistently makes sense, is fun, and keeps us coming back for more'. This reflects his mission to foster the growth of what he calls 'humanist technologists' – people who are capable of articulating future culture through informed understanding of the technologies they use.

John has picked up a number of highly prestigious awards including the National Design Award in 2001, (the USA's highest honour for design), and Japan's highest career honour, the Mainichi Design Prize.

Pioneers can't please everyone. John was introduced to a stranger at a private party who said, 'John Maeda? I know you. I hate you.' He continued, 'I can't stand what you stand for'. But the feeling is somewhat mutual. John was once asked, 'What do you hate most?' To which he replied, 'Society's resistance to change'.

**Story Musgrave** ***Astronaut***

When you discover how much and how many different things Dr Story Musgrave has achieved, you do wonder if this is all about one person.

To say that he's the only astronaut to have flown on all five space shuttles is a mere fraction of his astonishing story.

He's also a pilot, surgeon, mechanic, poet and philosopher.

By the age of 10, Story was operating and repairing tractor and farm machinery on the family's 1,000-acre dairy farm in Massachusetts. By the age of 16, he'd made his first solo flight.

His fascination with space began on the farm. From the age of three he would head for the surrounding forests, lie on his back and gaze at the stars.

His journey to being one of NASA's most experienced astronauts, though, was anything but direct.

He left school before graduating and joined the US Marine Corps. Story returned to the USA, left the Marines and enrolled at Syracuse University to study for his first of an impressive list of degrees. He gained his Batchelor of Science degree in mathematics and statistics in 1958.

His next was a Masters Degree in business administration and computer sciences at California's UCLA.

A Batchelor of Arts degree in chemistry followed, and then he graduated with a Doctorate in medicine from Columbia University, New York in 1964. He completed a Master of Science degree in physiology and biophysics in 1966.

In 1967, Story was one of only 11 people selected from around 4,000 applicants to join NASA's astronaut corps. This reflected a shift in NASA's recruitment policy to include scientists, where previous candidates had been selected from the ranks of military test pilots.

It would be 16 years before Story flew his first space mission. This was aboard the maiden voyage of the Space Shuttle Challenger in 1983. During this mission he and Don Peterson became the first astronauts to perform a space walk from the Shuttle programme. Overlapping Story's NASA career, between 1967 and 1989, he continued to work as a part-time trauma surgeon. He also completed a Master of Arts degree in literature.

Story's fifth and penultimate space mission involved a record five spacewalks, three of which were undertaken by Story. The mission was to carry out highly complex repair and maintenance of the Hubble Space Telescope. In all, Story recorded 1,281 hours 59 minutes and 22 seconds in space, and around 25 million miles in orbit.

Story has also flown nearly 18,000 hours in 160 different types of civilian and military aircraft as pilot, instructor and acrobatics specialist. He's made more than 600 private parachute jumps.

Oh, and then there are his hobbies: gardening, chess, reading, microcomputers, gliding, scuba diving, photography, literary criticism and running.

The average renaissance man looks idle by comparison.

**Julie Taymor** ***Director etc.***

You can't pigeon-hole Julie Taymor, and she's delighted about it.

She's an opera, film and theatre director and a designer, and she's enjoyed critical acclaim, and awards, across these disciplines.

The first flowering of her varied talents emerged in Julie's backyard in Newton, Massachusetts where, by the age of seven, she was producing plays.

Born in 1952, Julie started acting at the Children's Theater in Boston when she was 11.

It wasn't long before she started gathering theatrical and cultural influences from much further afield.

Her interest in Asian theatre began when she travelled to India and Sri Lanka, aged 15.

The following year, she was studying mime in Paris with the legendary Jacques le Coq.

Her travels – and influence gathering – continued after she graduated from Oberlin College in Ohio in 1974, and won a Watson Fellowship.

She headed for Eastern Europe and on to Asia. She stayed in Indonesia for four years. She was captivated by an environment where theatre wasn't an exclusive event for an elite audience, but very much a part of everyday life.

She set up the Teatr Loh, an international performing troupe in Bali, before heading to New York in 1974.

The richness of the cultures she absorbed bubbles to the surface in the fantastic visual feasts she prepares for her audiences. Understandably, and justifiably, she enjoys a reputation for this around the world.

Two of the highlights of her adventures in opera are her direction of Strauss's *Salome* for the Kirov Opera in St Petersburg, and of Mozart's *The Magic Flute* for the Maggio Musicale in Florence. Her production of *The Magic Flute* is now in repertoire at the Metropolitan Opera, New York.

Perhaps her greatest triumph, so far, as a stage director has been *The Lion King*. Her credits for *The*

*Lion King* bear testament to her numerous outstanding talents.

Director, lyricist (for the song 'Endless Night'), costume designer, co-mask designer, co-puppet designer. In addition, Julie won a Tony for Best Direction of a Musical and a second Tony for Best Costume Design.

Her award-winning extends to her film direction, too. *Frida*, Julie's 2002 movie, based on the life of artist Frida Kahlo, received six Oscar nominations. Two converted into Oscars for Best Original Score and Best Make Up.

Her previous film, *Titus*, in 1999, starring Sir Anthony Hopkins, received very positive reviews.

Her latest film, completed in 2006 and called *Across The Universe*, stars Bono and Evan Rachel Wood, and features the music of The Beatles.

Her latest project is the direction of a newly commissioned opera (quite a rare event), *Grendel*, which her partner, Elliot Goldenthral composed.

Many of Julie's productions, including *Titus*, have been accompanied by illustrated books. There's also one covering her rich and varied career, called *Julie Taymor: Playing with Fire – Theater, Opera, Film.*

Given her visual strengths, they really couldn't be anything other than illustrated.

**2005 Saatchi & Saatchi Award for World Changing Ideas**

**The Ideas**

Concrete Canvas · Peter Brewin/William Crawford SAATCHI & SAATCHI AWARD WINNER
Lens-Free Ophthalmoscope · Roger Armour EDWARD DE BONO MEDAL WINNER
Bio-Solar Energy Nanodevices · Marco Baldo/Shuguang Zhang
Frozen Ark Project · Dr Ann Clarke/Professor Bryan C. Clarke/Dame Anne McLaren
Jot a Dot · Quantum Technology
Optical Stretcher · Dr Jochen Guck/Dr Josef Käs
Photo-Form Tactile Graphics · Keith Carlson
Plantic® · Plantic Technologies
Splashpower · Pilgrim Beart/Lily Cheng/James Hay
Subvocal Speech Recognition · Dr Chuck Jorgensen
Wikipedia · Jimmy Wales

**The Judges**

**Chris Anderson** ***TED Curator***

Chris is a Brit who was born in a remote village in Pakistan in 1957. He spent his early years in Pakistan, India and Afghanistan (which at the time, he says, was beautiful).

His father worked as a missionary eye surgeon. His early education was at an American school in the Himalayan Mountains, transferring to boarding school in Bath, England, when he was 13.

Chris graduated in philosophy from Oxford University in 1978.

After training as a journalist, and becoming hooked on computing, Chris launched his own media company, Future Publishing, in 1985.

After seven highly successful years, Chris sold the business to Pearson and headed for the US in 1994 to try to do the same again.

Imaging Media took off and Chris re-merged it with Future, floating the new entity in London in June 1999.

A year of stock market glory followed. Two thousand people, $2 billion market capitalization, 130 magazines. Then came the dot-com bomb.

Chris saw 95 per cent of the value he thought he'd built evaporate.

He turned his focus to the Sapling Foundation, which he'd founded and funded well before the collapse, and its prize asset, TED.

TED – Technology, Entertainment, Design – is an annual conference that has been described as 'the ultimate brain spa', and a 'four-day journey into the future in the company of those who are creating it'.

In 2003, Chris led a TED group to brainstorm a new type of award. What emerged was the TED Prize, which grants three wishes each year to three world-changing individuals.

Chris lives in Manhattan and divides his time between the TED Conference, the Sapling Foundation and the TED Prize.

He considers himself to have one of the world's most satisfying jobs.

At the same time, he believes, 'it's a terrifying thought that it's not just good ideas which change the world. Bad ones do too. Like the idea that my race or religion is better than yours, or the idea that someone else will fix the problem. Our survival depends on the raging battle in the global ecosystem of brains between ideas that inspire and those that destroy.'

**Edward de Bono** ***Lateral Thinker*** *(page 237)*

**Philip Glass** ***Composer***

It's been said of Philip Glass that 'no musician since Stravinsky has had so great an impact on the sound of music of his own time'.

One of a number of composers described as minimalist, Philip Glass came to his unique musical place by a far-ranging, indirect route.

The first music he discovered came from his father's radio repair shop in Baltimore. These were the records that weren't selling and they tended to be the more obscure classics, Shostakovich symphonies, for example.

Glass began to take music lessons when he was six, in 1943. He worked at the flute between the ages of eight and 15.

He successfully applied to the University of Chicago during his second year in high school, and moved there to major in mathematics and philosophy. When he wasn't supporting himself with part-time jobs as a waiter or loading aircraft, he practised the piano.

Glass graduated at 19 and moved to New York to attend the Julliard School. He was drawn to maverick composers such as Charles Ives, Harry Partch and Moondog.

He moved to Paris in 1960 to study under Nadia Boulanger. It was here that Glass discovered the techniques of Indian music by working on Ravi Shankar's ragas.

He went on to research music in North Africa, India and the Himalayas before returning to New York and applying what he'd learned to his own composing.

In the New York of the early 1960s, minimalism in painting and sculpture influenced the composers in the city such as Glass and Steve Reich. In 1968 Philip Glass formed the ensemble he needed to perform the music he was composing. This was minimalist music performed at rock music volume.

In 1975 Glass found himself without a record contract. He started working with Robert Wilson on the first of three operas about historical figures 'who changed the course of world events'.

*Einstein On The Beach* is a five-hour epic and is recognized as a landmark in twentieth-century music and theatre.

Opera is just one musical genre explored with great success by Philip Glass.

Film scores, for instance, have included the Qatsi film trilogy, *Koyaanisqatsi*, *Powaqqatsi* and *Naqoyqatsi*, and *The Thin Blue Line*. He received an Oscar nomination for the score for Martin Scorcese's *Kundun* and for Stephen Daldry's *The Hours*, for which he also received Golden Globe and Grammy Award nominations. The score for *The Truman Show* won him a Golden Globe Award.

And then there's the orchestral works, chamber works, dance and theatre compositions. He identified the motivation to find his own musical voice when he was in Paris. 'Playing second fiddle to Stockhausen didn't seem like a lot of fun – there didn't seem to be any need to write any more of that kind of music. The only thing was to start somewhere else.'

**Baz Luhrmann** ***Filmmaker***

Baz Luhrmann was born in New South Wales, Australia, on 17 September 1962.

In the tiny rural settlement of Heron's Creek where he grew up, his father ran a petrol station and a movie theatre.

The theatre captured Baz's imagination and inspired his future. He became enthralled by the power of storytelling.

He is the founder and director of Bazmark Inq. and its subsidiaries, Bazmark Live and Bazmark Music. Bazmark's motto is revealing, 'A Life Lived In Fear Is A Life Half-Lived'.

Baz is an innovator and has enjoyed success across a wide range of projects in film, theatre, music and events.

He's staged Puccini's *La Boheme* on Broadway. He's created a concept album, *Something For Everybody*, featuring 'Everybody's Free to Wear Sunscreen' which entered the charts at No. 1 in the UK.

He was an adviser to Prime Minister Paul Keating during his successful 1993 re-election campaign in Australia.

He staged his interpretation of Benjamin Britten's version of *A Midsummer Night's Dream*, set in colonial India, for the Australian Opera.

This is Baz's view on music. 'Music touches and unites us in a way words cannot describe. Music is there when words fail us.'

Baz's deliberate use of counter-contemporary music for his Oscar-winning film *Moulin Rouge!* raised some critical eyebrows but was met with huge popular acclaim.

*Moulin Rouge!* followed *Strictly Ballroom* and *William Shakespeare's Romeo + Juliet* to form Baz Luhrmann's Red Curtain Trilogy.

Baz explains his Red Curtain style of filmmaking in this way. 'Today's audience is very aware that cinematic reality is just a manipulative device. The Red Curtain style simply makes a new deal with the audience. It pulls back the curtain and says "we, the storytellers are here."'

**Lou Reed** ***Musician***

The word 'legend' is in danger of being overused and its meaning devalued. It should be reserved for people like Lou Reed.

In his long career, he's tackled taboos and broken down barriers. He's both fearless and, to many, mainly journalists, highly irascible.

He's the songwriter and guitarist who teamed up, in 1965, with classically trained violinist and pianist John Cale, bass and guitar player Sterling Morrison, drummer Maureen Tucker and the enigmatic singer, Nico, to form The Velvet Underground. The Velvets were the shocking antidote to the cheerful fluffiness of mid-sixties Moon-in-June pop music.

Their songs like 'Heroin' took teenage music to places where the parents didn't want it to go. Reed's music was a little too reality-based for their liking.

The Velvets' association with Andy Warhol elevated the band to a different artistic model. They introduced and popularized mixed-media happenings, using projected film, strobe lighting and dancers in their live shows.

Six years and four albums later, Lou Reed started his solo career but continued to be a pioneer. An admiring David Bowie said, 'The nature of his lyric writing had been hitherto unknown in rock. He supplied us with the street and the landscape, and we peopled it.'

Bowie produced Lou Reed's 1972 album, *Transformer*, which for many is a truly seminal piece of work.

One of its tracks, 'Walk on the Wild Side', deals with trans-sexuality and was a top 20 hit in the US and made the top 10 in the UK.

The implication is that the whole popular music landscape had been changed by now, so fringe subject matter fitted the mainstream.

Reed's contribution to this transformation can't be overestimated. His album's title could not have been more apposite.

In 1987, Lou was reunited with John Cale at Andy Warhol's funeral. They were inspired to collaborate on a musical biography of Warhol. *Songs for Drella*, was a reference to Warhol's nickname 'Drella', a combination of Dracula and Cinderella.

In the late 1990s Lou was inducted into the Rock and Roll Hall of Fame, made a Chevalier Commander of Arts and Letters by the French government, and received the Hero Award from the National Academy of Recording Arts and Sciences. Does this mean Lou Reed has joined the establishment? Probably not.

**Sue Savage-Rumbaugh** ***Primatologist***

Dr Sue Savage-Rumbaugh is a primatologist, experimental psychologist and one of the world's leading ape language researchers.

Her interest in ape language was triggered over 30 years ago when she was studying child development. During a psychology seminar, she was introduced to a chimpanzee which had been taught deaf sign language.

This prompted Sue to set out to find how we humans came to be what we are – of which language is just one aspect.

She says, 'Language is the home of the mind. It's humankind's tool. We apply it to everything. But research is bound by the current constraints of our science. Traditional scientific approaches do not work when studying relationships. No interaction ever repeats itself, not between humans and chimpanzees, not between two humans.'

After 23 years at the highly regarded Language Research Center at Georgia State University, Sue moved in 2005 to the Great Ape Trust in Des Moines, Iowa. Here, Sue has a purpose-built facility to continue her work.

At the heart of her research are eight bonobos – pygmy chimpanzees. Banobos and humans share 98.4 per cent of the same DNA.

The best known of the bonobos is Kanzi. He is the hero of Sue's 1995 book, *Kanzi: The Ape at the Brink of the Human Mind.*

The key to communication between humans and Kanzi and the other bonobos in Sue's community are lexigrams or symbols for English words. In practice, the symbol itself is unimportant. What is essential is consistency.

With 400 lexigrams in operation, it's possible for quite complex conversations to take place. Although a bonobo's vocal tract is designed for vowels and not consonants, Sue believes her bonobos can utter certain recognizable English words. The bonobos are also using some English words to speak to each other.

Kanzi, for example, has a vocabulary of several thousand words, and his cross-species communication skills have extended to music, having jammed on his keyboard with the likes of Paul McCartney and Peter Gabriel.

Sue says her bonobos 'have learned to become more human, and I've learned how to become more bonobo'.

**Oliviero Toscani** ***Photographer***

Oliviero Toscani breeds horses and produces olive oil in Tuscany.

He's also the photographer who created perhaps the most controversial advertising campaign there's ever been, for the Italian fashion brand, Benetton.

His images, including a nun and a Catholic priest kissing, a dying AIDS patient, a death row prisoner and a newly born baby with umbilical cord still intact, were provocative and impactful.

Not everyone responded positively to this unexpected issue-highlighting work. A common response was the question, 'What has it got to do with fashion?'

Dominique Auginot, the photographer and President of Lux Modernis, a French advertising company, was more pointed. 'Toscani is on another planet. I think his work is sick and unhealthy.'

No amount of controversy, however, can take away from Toscani's genius as a photographer.

His work has been exhibited at the Biennale of Venice, the Triennale of Milan, in São Paolo, Lausanne, Mexico City, Helsinki, Rome and more. He's also won a whole clutch of prestigious awards.

Oliviero was born in Milan in 1942, the son of a photojournalist. He studied design and photography at the Hochschule für Gestaltung in Zurich.

As a fashion photographer his work has appeared in pretty much every stylish magazine worthy of the description.

In 1990, emerging from the United Colors of Benetton stable came the magazine *Colors*, which he created in collaboration with the late and legendary designer Tibor Kalman (see page 239), after interviewing and rejecting many other candidates.

The *Colors*' tag line 'a magazine about the rest of the world' indicated its commitment to the multiculturalism Toscani had explored in his Benetton ad campaigns.

In 1993, Toscani founded Fabrica, the international centre for research in the arts of modern communications. The centre's ultimate goal is to uncover new forms of visual and aural communication and to realize that creativity is not an abstract phenomenon. It is an idea that becomes a reality.

Reassuringly, there's no sign of Toscani losing his ability to provoke.

His recent ad campaign for the men's clothing brand 'Ra-Re', had images of men engaged in homosexual behaviour. The reaction was predictable.

Toscani seems to be incapable of resisting any opportunity to be controversial.

His response to a journalist's question, 'What project has given you the most satisfaction?' was 'Masturbation'.

## Picture credits

Nick Beare 226
David Becker 169
Keith Carlson 170, 172, 173tr, 173mr, 173br
Corbis 112 (Bill Ross)
Design that Matters 216
Digit Wireless, Inc. 96
© FanWing Ltd 87 (Ernesto Tross), 88–89 (Ernesto Tross)
Getty Images 20–21 (Alan R. Moller), 38–39 (AFP)
George Grall 173ml
© Hariri & Hariri Architecture Engineering/Mark Borstelmann, LCD Planar Optics 118–119, 120
Integrated Medical Systems and their partners and collaborators 146–147
iStockphoto.com 47 (Suzannah Skelton)
LG Electronics 97
LUTW 230
Courtesy of The Long Now Foundation 78 (Rolfe Horne), 79 (Rolfe Horne)
NASA 60 (Ames Research Center), 173tl (Cassini Spacecraft)
© Natural History Museum, London 25, 26, 28
Anthony Oliver 111, 115, 123, 126–127, 142, 159, 211, 213
Plantic® 43, 45
© Splashpower Ltd 130
Bill Stelle 173bl
Elspeth Twist 224
Jan van der Crabben 75

## Index


Adaptive Eyecare Company 212
Aldrin, Buzz 9, 50, 68, 136, 236
alphabets 63, 64, 66
Alzheimer, Alois 152
Anderson, Chris 251
Anderson, Laurie 206, 237, 243
Armour, Roger 225, 251
Artificial Vision for the Blind 201–205, 246
B Consultants 105
Baldo, Marco 46, 251
Barker, Tom 105, 108, 246
batteries 12, 122, 124, 129, 131, 162, 217, 229, 232
Baylis, Trevor 125, 136, 242
Beart, Pilgrim 251
Bellenson, Joel 110, 242
Biochip Artificial Vision Project 196–200, 236
Biomechanical Transducer/Electric Shoe 125–128, 242
BioMuse 184–187, 236
Bio-Solar Energy Nanodevices 46–49, 251
blindness 168, 171, 174, 178, 179, 188, 190, 191, 196, 200, 201, 205, 227
Bloor, David 133
Bono, Edward de 68, 136, 206, 237–238, 242, 246, 251
Braille 171, 188, 190, 191, 205
Braille, Louis 188, 191
Brewin, Peter 31, 251
Burke, James 136, 238
Byrne, David 243, 246
Carlson, Keith 171, 251
Champollion, Jean-François 81
Cheng, Lily 251
Clarke, Ann 29, 251
Clarke, Bryan C. 29, 251
Concrete Canvas 30–35, 251
Cooper, Martin 116
Cowley, R. Adams 148
Craig, Ashley 194, 246
Crampin, Stuart 37, 40, 41, 246
Crawford, William 31, 251
Critical Mass Communicator 219–222, 236
Crossed Beam Display 91–94, 236
Davies, Paul 242
De Bono Medal 68, 82, 240, 242, 246, 251
Design that Matters 215, 246
diabetes 227
DigiScents 110–113, 242
Digital Building Blocks 117–121, 242
diseases 27, 149, 151, 152, 154, 156, 157, 195, 196, 223, 225, 227, 232
DNA 27, 29, 49, 248
Dobelle, William 201, 205, 246
Downing, Elizabeth 91, 236
earthquakes 34, 36, 37, 40
education 11, 117, 140, 215, 218, 232
electricity 11, 116, 128, 186, 217, 219, 228, 229
endoscopic surgery 163
energy 44, 46, 48, 49, 122, 124, 125, 128, 228
Eno, Brian 242–243, 244, 246
extinction 24, 27
FanWing 86–90, 246
Faraday, Michael 116
Fastap technology 95, 98
Feynman, Richard 37
Frozen Ark Project 24–29, 251
Galvani, Luigi 131
garbology 44
genetics 27, 29
geology 37, 41
Gibson, William 136, 238–239
glass 121
Glass, Philip 251–252
Global Earthquake Monitoring System (GEMS) 36–41, 246
Golden Hour, the 144, 148
Granger, Richard 152, 246
Guck, Jochen 157, 251
Hadid, Zaha 105
Hanson, Matthew 145
Hariri, Gisue 117, 120, 242
Hariri, Mojgan 117, 242
Hay, James 251
Hillis, W. Daniel 246–247
honeycomb 108, 109
Humayun, Mark 196, 197, 200, 236
Hutton, James 41
Iddan, Gavriel 162, 242
Integrated Medical Systems Inc. 145, 242
Intelligent Mobile Phone Charger 114–116, 246
Internet 100, 102, 103, 117, 206, 248
Irvine-Halliday, Dave 164, 228, 229, 232, 234, 246
James, Barry 125, 242
Jorgensen, Chuck 59, 251
Jot a Dot 188–191, 251
Juan, Eugene de 197, 236
judges 8, 14, 68, 236, 242, 246, 251
Kalman, Tibor 68, 239–240, 254
Käs, Josef 157, 251
KASPA 174–179, 234, 236
Kay, Leslie 174, 178, 179, 234, 236
Kelly, Kevin 136, 243–244
Kinkajou Projector 215–218, 246
Kitatani, Kenji 247–248
Knapp, Benjamin 184, 236
language/s 54, 56, 57, 58, 60, 72, 73, 76, 77, 79, 80, 140, 191
Lanier, Jaron 219, 220, 236
lasers 91, 153 157
Lee, Ji 50, 63, 64, 236
Lens-Free Ophthalmoscope 223–227, 251
Levy, David 13, 95, 236
light 91, 157, 196, 204, 229
lighting 11, 12, 105, 164, 228, 229, 232
Light Up The World Foundation 164, 228–233, 234, 246
Lipman, Adi 54, 236
Lipman, Aharon 54, 56, 236
literacy 218, 220
Long Now Foundation, The 77–78, 242, 243, 244, 247
Loomis, Mahlon 116
LSTAT (Life Support for Trauma & Transport) 144–148, 242
Luhrmann, Baz 252
Lussey, David 132, 134, 242
Maeda, John 248
Maes, Pattie 244
McLaren, Anne 29, 251
Meron, Gavriel 162, 242
Mills, Tim 162, 242
Mind Switch 192–195, 246
Minsky, Marvin 244, 246
mobile phones 12, 62, 98, 108, 114, 116, 122, 124, 131
chargers 114, 116, 122, 124, 131
Monteith, John 125, 242
Murdoch, Lachlan 240
Musgrave, Story 248–249
NASA 59, 62, 92, 132, 134, 236, 237, 247, 249
NeuroGraph Diagnostic Aid System 149–152, 246
Nordström, Kjell A. 244–245
O'Connor, Richard 246
ophthalmoscope 223, 225
Optical Stretcher 153–157, 251
Optyse LFO 225, 227
Ovellhaus, Gerhard 140, 242
Pappas, Ted 73
paralysis 180, 181
Peebles, Patrick 86, 90, 246
peptide detergents 49
Peratech 134, 135
Perkins Brailler 190
Peterson, Richard 140, 242
Photo-Form Tactile Graphics 168–173, 251
photosynthesis 46, 48
Plantic® 42–45, 251
Prestero, Timothy 217
Quadkey 95–99, 236
Quantum Technology 191, 251
Quantum Tunnelling Composites (QTC) 132–135, 242
Quicktionary 54–58, 236
Rabischong, Pierre 181, 182, 246
Rathje, William L. 44
Reed, Lou 206, 237, 252–253
Ridderstråle, Jonas 244, 245
Roberts, Kevin 82, 136, 164
Rosetta Disk 77–81, 242, 244
Rosetta Stone 80, 81
Sanger, Larry 72, 73
Savage-Rumbaugh, Sue 253
Seismic Detection of Tornadoes 18–23, 236
Self-Adjustable Spectacles 82, 210–214, 236
Silver, Joshua 82, 210, 212, 214, 236
Slate and Stylus 188
SmartSlab 104–109, 246
Smith, Dexster 110, 242
Solar-Powered Mobile Phone Charger 122–124, 246
Sonic Torch 178
SplashPad 131
Splashpower 129–131, 251
Sprint Integrated On-Demand Network 100–103, 236
Stand Up And Walk Project 180–183, 246
Stokes, Paul 114, 246
Subvocal Speech Recognition 59–62, 251
Sun, the 7, 37, 124
Swain, Paul 162, 242
Tatom, Frank 19, 23, 236
Taymor, Julie 249–250
television 36, 120, 192, 194, 195
Telus Mobility 98
3-D Technology Laboratories 92
tornadoes 18, 19, 22, 23
Toscani, Oliviero 239, 240, 253–254
UN 24, 30, 34, 145, 218
Univers Revolved 63–67, 236
Universal Pill Marking System 140–143, 242
virtual reality 219, 222
visual cortex 201, 204, 205
VorTek 23
VPL Research 219
Wales, Jimmy 72, 73, 206, 251
Wikipedia 72–76, 206, 251
Wireless Capsule Endoscope (Pillcam) 158–163, 242
WizCom Technologies Ltd 56, 57
WHO 145, 188, 227
Wurman, Richard Saul 240–241
Young, Thomas 81
Zhang, Shuguang 46, 49, 251